HISTORIC FIRES OF FALL RIVER

STEFANI KOOREY, PhD

THE History PRESS

Published by The History Press
Charleston, SC
www.historypress.net

Front and back covers: All images from the Leslie Jones Collection.

First published 2016

Manufactured in the United States

ISBN 978.1.46711.924.5

Library of Congress Control Number: 2015960002

Notice: The information in this book is true and complete to the best of our knowledge. It is offered without guarantee on the part of the author or The History Press. The author and The History Press disclaim all liability in connection with the use of this book.

To Harry Widdows, who deserves much more than this dedication.
A longer life would have been my choice. I will miss you always.

CONTENTS

FOREWORD

Fall River is historically known as the City of Hills, Mills, and Pork Pies. Former fire chief Louis Shea had a different view, famously describing Fall River as a city that was "built to burn."

Located at the confluence of the Taunton and Quequechan Rivers, Fall River was once the cotton textile manufacturing center of the United States, second only to Manchester, England, in the world. This industry brought waves of immigrants to the city—English, Irish, French Canadians and Portuguese—all flocking to Fall River during the nineteenth and twentieth centuries to work in the mills. These workers needed cheap, affordable housing, and so the "three-decker" was born. These apartment houses usually had three floors with porches on each level. The wood framing of these structures consisted of two-by-fours running from the roofline down to the foundation with a ribbon board at each level where the floor joists were attached. Oftentimes, the walls of these buildings were never fire-stopped, so a fire that started in the cellar could quickly communicate to the attic due to its balloon framing. These buildings were jammed into neighborhoods so closely that people could not stretch their arms out between them. The mills that the people worked in were located in these neighborhoods, as well. They were and are huge buildings of granite or brick with oil-soaked floors from the spinning and weaving machines. When they caught fire, they became roaring infernos.

There are only three ways to extinguish a fire: cut the oxygen to the flame, cool it below its ignition temperature or let it burn itself out. A fire is a force of nature and is a living, breathing thing, like a hurricane or tornado. It needs oxygen to live and will always follow the path of least resistance. In mill buildings, where there are great open areas that are hundreds of feet in length, there is little to stem the flow of the flame once it has begun.

I was a firefighter in the city of Fall River for thirty-three years, retiring in 2009 as a district fire chief. During this time, I fought many fires, both large and small—Kerr Mill, Arlans (Richard Borden mill complex) and Notre Dame de Lourdes church, among others. The worst fire to me, however, was one that occurred on December 28, 2000, in the Corky Row section of the city. As with other conflagrations in Fall River, there was an icy, cold wind, this time recorded at forty miles per hour, from the southwest. We had a fire call at 2:00 a.m. that reported a house fire on Fourth Street. The temperature was nine degrees. As we approached the fire, we saw four vehicles ablaze on the street from the flames that had traveled to the next flammable object. It was so cold that we had two frozen hydrants and had to lay hundreds of feet of hose to get to a flowing hydrant. By the time of our arrival, a six-family house, a three-decker and a three-family house were all on fire. Tragically, one young mother died and twenty-five people were displaced. Our firefighting efforts that night saved the neighborhood, and even though everything was covered in a thick coat of ice, including ourselves, we stuck it out until we were sure the danger had passed for the surrounding structures. We almost lost an entire section of the city that night.

Another horrible fire that I recall also resulted in a loss of life, this time four people were killed in the Holy Ghost Club fire of 2006. Located on County Street, the club had installed a religious shrine, using cotton batting on the floor, walls and ceiling. Illumination was by candles, and when one of them fell over, the building caught fire quickly. When there is any loss of life, firefighters take it very personally.

The conflagration of Notre Dame de Lourdes church in the Flint area of Fall River in 1982 was one of the most significant fires in my tenure as firefighter. After lunch, on Tuesday, May 11, and for five full hours, the Fall River Fire Department and companies from Westport, Dartmouth, New Bedford and as far away as Boston worked to contain this all-consuming blaze. By the time my company arrived, multiple buildings were on fire on the southwest side of the church, toward Pleasant Street, a major business district in this part of the city. We set up water cannons on the east side, directing the water onto the fire. The fire was so large that it created its own weather and wind patterns, a firestorm, with fire cresting over the ladders and cannons, backward and in a circular motion. At ground level, the air was rushing in a direction toward the fire that it was creating—opposite of the wind direction of the environment. Embers from the fire traveled quite a distance, and we put out brush fires near the highway in the bull rushes of the Quequechan River well into the night. The headlines in the newspaper said it all: "The Day a City Wept."

We never know what we are going to find when we are called to a fire. We may get a call for a stove fire or mattress fire but have no certain knowledge about what is actually happening and what we will find when we arrive at the scene. We don't even have a firm idea as to how many people actually inhabit a building. There could be—and it has been my experience to find—unlawful apartments in the attics and cellars of a house that is on fire, as well as rooms filled with all kinds of illegal and/or flammable containers and accelerants—gasoline, kerosene, oxygen tanks, acetylene, butane, fireworks, gunpowder, et cetera. There is lots of unpredictability. And with unpredictability comes danger, for both citizens of the city and the firefighters who will risk their lives to save life and property, regardless of type of emergency, weather or hour.

Over the years, Chief Shea's description of Fall River has unfortunately borne out. *Historic Fires of Fall River* will go into greater detail about the conflagrations, mill fires and other fire disasters, both natural and man-made, that Fall River has suffered throughout its history. The fire department will continue to protect the lives and property of the citizens of Fall River, and we have been fortunate to have such a dedicated and professional group of people to protect the city from the inevitable ravages of fire.

David Jennings
Retired District Fire Chief
Fall River Fire Department

ACKNOWLEDGEMENTS

As with any nonfiction work, the process of creation is never a solo endeavor. There are librarians, historians, experts, organizations and others who all contribute in a pronounced way to the final product. This book is no different.

The Fall River Historical Society has been most gracious in allowing me, once again, access to the collection of Philip D. Borden, former city engineer for Fall River—a collection containing literally hundreds of images of the city from every conceivable location, even its nooks and crannies and ugly spots. The society granted me permission to select from these images, including ones from the Granite Mill Fire of 1872, illustrations that are too expensive to otherwise obtain. Curator Michael Martins and assistant curator Dennis Binette deserve a great deal of thanks for their assistance.

Many of the images contained herein were taken by a news photographer whose work is on digital display with the site DigitalCommonwealth.com, from the collection of the Boston Public Library. They have proven to be the most dramatic images of fires in Fall River that I discovered, and a big thank-you goes to the family of Leslie Jones and the Boston Public Library for permission to reprint them here. Remarkably, Jones's images span two fires contained in this book (the Steiger Store Fire of 1916 and the Great Fire of 1928). I know you will find these photographs as compelling as I did when first I laid eyes on them.

Dave Jennings, former division chief of the Fall River Fire Department, was instrumental in making this work possible, and I thank him for his assistance in helping this plebe understand firefighting techniques and useful jargon, including the proper naming of apparatus. I don't think I quite understood what a firefighter actually does until he explained it to me. Dave is the curator of the Lafayette-Durfee House in Fall River and entertains while teaching children of all ages the history of the American Revolution, the Civil War, pirate life and colonial America.

Philip T. Silvia, PhD, was so supportive of this project and quite excited about my involvement that he regularly offered new finds to me when he ran across them in his Fall River history collection. His massive two-volume dissertation, "The Spindle City: Labor, Politics, and Religion in Fall River, Massachusetts, 1870–1905," was an important text for me to read and very much helped me form my understanding of mill politics, union history and the cultural configuration of Fall River in ways that I had never before grasped. For these reasons, Phil's work was seminal to the creation of this book.

The staff and management of the Fall River Public Library are owed a debt of gratitude for their patient assistance with this project. The Fall River Room at the main branch was my first stop in the hunt for fire articles and smaller works. I would like to thank library director Laurel Clark and reference librarian Kate Kulpa for their assistance.

A thanks also goes to my acquisitions editor, Karmen Q. Cook, who is always supportive and an expert on keeping me on task to meet deadlines. Your kindness has not gone unnoticed. Julia Turner, production editor, likewise made the editorial process a wicked good experience.

On a personal note, I would like to thank those friends who once again put up with my hibernation during the writing of this book. They include Ian Garrett, Dipti Lincoln, Leonard Rebello, William Pavao, Kristee Bates, Joe Carvalho, Everett Castro, Barry French and Onsloe Brimbau.

A massive thank-you goes to my sweetie, Michael Brimbau, who not only created the physical space for me to write in but also supported me emotionally and personally through this process.

INTRODUCTION

This City now doth, like a garment, wear
The beauty of the morning; silent, bare,
Ships, towers, domes, theatres and temples lie
Open unto the fields, and to the sky;
All bright and glittering in the smokeless air.
—William Wordsworth

There is a haunting black-and-white image, taken by news photographer Leslie Jones on the morning after the Great Fire of 1928. In the foreground, from left to right, are four heavy-coated firemen, each on a different elevation of a massive ice hummock. Hazy smoldering ruins of some once enormous structure dominate the background. The men are dwarfed by their environment and appear like tiny dark figures in a giant's ice world. The ghostly skeletal remains of the burned structures appear *super*natural. This destruction must surely be the result of a wartime bombing raid.

As two of the men stare at the scene in front of them, seemingly transfixed by the enormity of the sight, two others are in ascent, following a fire hose line that snakes and cuts the ice-white ground ever upward. Everything is covered in a thick blanket of ice, the aftermath of fighting an intense blaze in zero-degree temperatures. This world is a dangerous place and in a delicate balance. We are overwhelmed by knowing that the remaining isolated pieces of once grand buildings will all have to be demolished and this debris field cleared of its tons of wood, metal and granite. And yet the men are frozen in time, as are the millions of gallons of water that were used to fight this

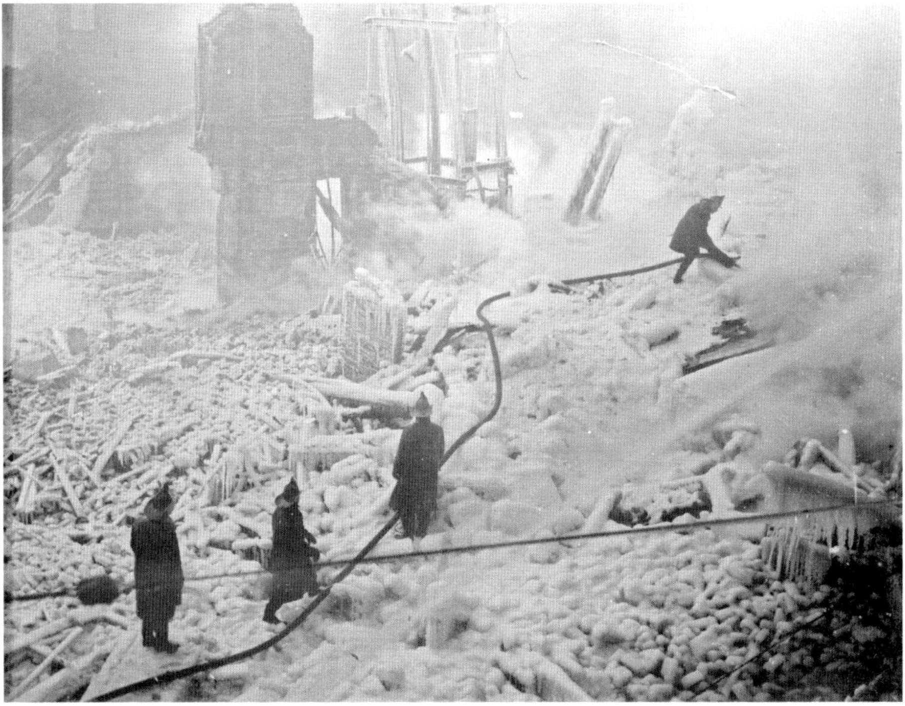

"Fall River Ruins," 1928. *Boston Public Library, Leslie Jones Collection.*

conflagration. This, the worst fire in Fall River's history, will forever remain seared in this city's collective memory.

Tragically, each succeeding generation has also had a substantial fire to contend with. From 1813 (before the city was incorporated) to 2015, Fall River has suffered through over ninety-five significant fires and many full-blown conflagrations. Amazingly, there has been relatively little loss of life to report, with the exception of some horrific house fires and the Granite Mill Fire of 1874, where twenty-three people, mostly women and children, died and over thirty others were maimed, burned or crippled from the flames or from jumping from the top attic floor, seventy feet to the ground.

Why so many fires in Fall River? New Bedford and Lowell both have mills, built during the same industrial expansion eras, but they don't have nearly the number of big fires. According to experts, it comes down to location. Fall River is located on the west side of a hill, a giant batholithic granite ledge formed eons ago in the ice age. The North Watuppa Pond, the second-largest naturally occurring body of water in the state and the source of the city's water supply, borders the city proper on the east; the Taunton River and

Mount Hope Bay form the boundary to the west. The winds are generally westerly and are at times quite gusty. Wind in Fall River has always proved to be an ally of fire.

The aging mills in Fall River, and there are many, are mostly over one hundred years old and were built for cotton manufacturing work, with multiple floors that are essentially 100- to 250-foot-long rooms with 15-foot-high ceilings and floors made of wide pine planks that have absorbed the oil dripped and spilled from machinery over the decades. When a fire occurred in a working mill in the past, usually the result of a friction spark that ignited cotton lint, there was little chance of stopping it. Firefighters would attempt to put out the blaze, but more often than not, they devoted most of their resources to ensuring that the fire did not spread to the surrounding businesses and residences. Fires in Fall River tend to create their own weather. As one child put it during the Kerr Mill blaze of 1987, "It was a tornado of fire."

Later mill fires, which occurred after the cotton manufacturing era, were also uncontrollable, but this time due to the way the structures had been refitted to accommodate multiple companies and concerns. "The old mill buildings were very well constructed," said Lieutenant Carlos Pacheco, head of fire prevention for the New Bedford Fire Department. "The floors are made of 3-inch planks in a tongue-in-groove style. They are very, very good fire stops. They're thick, and they take so long to burn."

But this "slicing" up of the mills to fit new uses and ways of working produced new dangers. When an old sturdy granite mill is converted to a dozen different factories, the walls are cut and reshaped, making for all kinds of voids and air spaces, producing vertical holes like chimneys or elevator shafts. When a fire starts in the cellar, and many mill fires do, it will spread very quickly up to the top floor, and total destruction is ensured. It is what every firefighter in Fall River fears.

Perhaps fear is the wrong word here. Firefighters don't "fear" fire; they fear the possibility of a massive conflagration and the effects that this will have on the community and local residents living near the blaze. They know that, sadly, Fall River is "built to burn." As one firefighter told me, whenever there is a loss of life or a fire destroys another building, they take it personally.

Aging tenement mill housing and triple-decker apartment buildings likewise are at risk of total destruction once a fire gets underway. The old balloon frame construction, without fire stops between floors, ensures the spread of fire from cellar to attic in minutes. Treated paper was the standard content of blown insulation, but the fire retardant in such materials leeches out with time, leaving behind an accident waiting to happen.

We must be mindful of the fact that a working mill was a business that employed hundreds or thousands of people. When a mill fire occurred, unemployment in the city would spike, leaving many on public assistance or dependent on the charity of churches or other organizations. The Firestone Fire of 1941 saw the loss of two thousand jobs and the Borden Mill fire, five hundred jobs. The Kerr Mill fire displaced 950 workers. In an instant, lives were forever changed. In each case, appeals would go out region wide as banks and businesses collected donated funds to offset the staggering loss incurred by the millworkers and their families. A fire is not just the loss of a structure. A fire in Fall River also has a very real human toll.

In Fall River, the smell of smoke is like a dog whistle—everyone takes notice. This ever-present knowledge that there *will be* another major fire in Fall River keeps the citizens on a kind of permanent alert. Neighborhoods are taking a more proactive approach to ensure that abandoned buildings and empty mills are monitored for trespassers. Even with high-tech alert systems in place and a fire department devoted to public safety, fires do happen, most recently in an empty school building, set by arsonists.

There are no easy answers or surefire fixes, pardon the pun, to this problem plaguing Fall River, Massachusetts. The most the city can do is what is being done—purchasing better safety equipment for firefighters and modern apparatus trucks as well as sustaining an appropriate force level. In a city where people live so very close to one another and industry and business coexist with residences, the best case scenario would be to have a well-equipped fire department with nothing to do.

CHRONOLOGICAL LIST OF FALL RIVER'S SIGNIFICANT FIRES

Bold entries signify a chapter in this book.

1813	Union Cotton Factory (Cook Pond)
October 26, 1821	Troy Cotton and Woolen Manufactory (friction fire that started in picker room), 1,800 spindles
1838	First Cotton Mill, original 1811 Wooden Mill, first textile mill in city, organized by Colonel Joseph Durfee
1838	Fourth Cotton Mill
July 2, 1843	**"The Big Fire" in center, loss of $526,485**
January 13, 1849	Steamboat Empire State, loss of $120,000
November 11, 1859	Fall River Rolling Mill, owned by Fall River Iron Works Company, loss of $50,000
November 12, 1867	Niantic Thread Mill (Division and Mulberry Streets), eighty operatives, loss of $50,000
December 5, 1867	Globe Print Works (boiler explosion), two hundred operatives, loss of $100,000
December 15, 1867	American Print Works, five hundred operatives, loss of $1,000,000
May 14, 1868	White Mill, built in 1827 (started in picker room), loss of $120,000
September 19, 1874	**Granite Mill, loss of $247,000, loss of life**

CHRONOLOGICAL LIST OF FALL RIVER'S SIGNIFICANT FIRES

November 2, 1875	Massasoit Cotton Mill, 175 operatives, loss of $150,000
June 29, 1876	American Linen Company Mill No. 2 (foot of Ferry Street), loss of $183,000
May 10, 1877	Kendrick Loom Harness Company (Pocasset Street), one death
November 17, 1877	Border City Mill No. 1, 450 operatives, loss of $450,000
November 20, 1878	Augustus Chace Thread Mill (Globe Street), loss of $50,000
October 20, 1882	Flint Mills, loss of $400,000
April 24, 1884	Sagamore Mill No. 1, five hundred operatives, loss of $500,000
January 12, 1885	Slade Mill, 350 operatives, loss of $200,000
March 19, 1886	City Hall, loss of $65,000
1888	Anawan Mill (struck by lightning and destroyed by fire)
July 16, 1890	Fourteen buildings burned from a fire started in John D. Crowley's junk shop, loss of $11,500
January 8, 1893	Troy Building, loss of $60,000
July 3, 1893	Cook, Borden & Co. Lumber Yard, loss of $13,000
November 13, 1893	Notre Dame Church (wooden structure), loss of $14,500
June 14, 1895	Langley's Loom Harness Factory (boiler explosion), four deaths, loss of $15,000
July 18, 1896	Granite Mill No. 3 cotton shed, one thousand bales of cotton and eight hundred bales of cloth, loss of $150,000
December 16, 1900	St. James Hotel
January 19, 1902	Globe Yarn Mill No. 1, seven hundred operatives, loss of $100,000
February 28, 1906	Granite Mill No. 2, storehouse and contents of 1,500 bales of Egyptian cotton (Robeson Street), loss of $125,000
September 15, 1907	Shove Mill No. 1, loss of $75,000
July 7, 1909	Fall River Golf Club

June 11, 1914	Six buildings on South Main, including Nelson Durfee Sawmill, AE Parker Welding Plant, a two-and-a-half-story tenement, two cottages and a stable
February 20, 1915	Troy Store, loss of $100,000
February 15, 1916	**Steiger Store and vicinity, loss of $1,000,000**
May 13, 1920	St. Anne's Church basement, loss of $80,000
July 24, 1920	New England Refining Company, loss of $20,000
December 26, 1921	Henry Lord School, loss of $250,000
April 5, 1923	Nicholas Building (Bedford and Sixth Streets), loss of $150,000
April 14, 1926	Baptist Temple, loss of $120,000
July 13, 1926	Street Railway Power House
January 12, 1927	Empire Theater, loss of $100,000
April 10, 1927	Technical High School and Winward Warehouse (Fourth Street), loss of $700,000
February 2, 1928	**Fall River Conflagration, loss of $10,000,000**
May 10, 1930	Sandy Beach Amusement Park, loss of $50,000
July 5, 1930	Five icehouses destroyed, loss of $25,000
October 27, 1931	Three-story wooden block (Pine and North Main Streets), loss of $20,000
July 4, 1932	Two icehouses on South Watuppa Pond, loss of $40,000
November 1, 1933	New England Terminal Co. (North Tiverton), loss of $800,000
November 21, 1933	Lis' Department Store, Borden Block, loss of $25,000
November 22, 1933	Waldorf Block, loss of $15,000
January 29, 1934	Merchants Mill (Fourteenth Street), one death from a falling wall, loss of $200,000
March 1, 1934	J. Desjardins Briar Pipe Co.
March 28, 1934	Davenport School, loss of $120,000
May 11, 1934	Magoni's Restaurant (Pocasset Street), loss $14,000
October 31, 1934	Staples Coal Co., Globe Dock, loss of $80,000
August 17, 1935	Wilbur's-on-the-Taunton, loss of $30,000

January 23, 1939	Zais Building (Edwards and Columbia Buildings), loss of $300,000
October 4, 1939	Midland Print Work on third floor of Davis Mill, loss of $60,000
February 6, 1940	Weetamoe Mill (WPA headquarters, owned by the city), eight hundred employees, loss of $450,000
February 26, 1940	Registry of Motor Vehicles and Goodyear Service (Union Mill Office Building), registry loss of $25,000, Goodyear loss of $7,000
March 25, 1940	Osborn Mill No. 2 (formerly known as the Montaup Mill), forty families driven from their homes, loss of $110,000
June 8, 1940	Estes Mill (on Sucker Brook), loss of $100,000
October 11, 1941	Firestone Rubber and Latex Company Fire, loss of $12,000,000
September 23, 1943	Hotel Mellen, loss of $200,000
October 16, 1946	Shove Mill, loss of $800,000
June 21, 1950	Baker Rubber Co.
January 22, 1953	Leigh Textile Co.
November 9, 1953	Borden School
September 16, 1960	Made-Rite Potato Chip Co.
November 5, 1960	Falls Mill No. 1 (Pocasset Street), loss of $1,000,000
August 15, 1961	Fall River Manufactory, Falls Mill No. 1 (Pocasset Street)
October 30, 1967	Bleachery Mill, loss of $250,000
1968	Monsignor Prevost High School (arson)
July 2, 1970	Nira Warehouse (Pond and Water Streets)
April 21, 1971	Church of the Ascension (Rock Street), loss of $15,000
July 2, 1971	St. Anthony of the Desert Church (Quequechan Street)
March 8, 1973	Former Firestone complex, Iron Works Cotton Mill No. 2, loss of $2 million
April 10, 1973	Sts. Peter and Paul Church
October 8, 1973	Weetamoe Yacht Club (vacant), loss of $10,000

February 18, 1974	Three stores in central business district of South Main Street (Shoe World, Wilbur's Dresses, vacant Thom McAn Shoes), loss of $110,000
November 19, 1975	Robeson School, loss of $350,000
February 26, 1978	Dwelly Elementary School, loss of $300,000
March 10, 1980	Pleasant Street fire, five deaths
November 5, 1981	Richard Borden Mill No. 1, loss of $5 million
May 11, 1982	**Notre Dame de Lourdes Church**
September 2, 1983	Unitarian Church (North Main Street), city's oldest church, loss of $2 million
January 12, 1987	Kerr Thread Mills Nos. 1, 2 and 3
April 21, 1991	St. Stanislaus Church (Rockland Street), loss of $500,000
Mid-1990s	Stevens Manufacturing Bleach House, once part of Bristol Finishing, built in 1892
1999	Chase Mill No. 2
June 14, 2006	Holy Ghost Club (County Street), four deaths, loss of $1 million
June 18, 2009	Oriental Chow Mein Noodle Company (rebuilt)
July 6, 2013	Sagres Portuguese Restaurant (rebuilt)
May 3, 2014	Mesa 21 restaurant (rebuilt)
August 22, 2015	Coughlin School (arson)

1

THE GREAT FIRE OF 1843

God of universal Providence for wise purposes, on that day, commissioned the elements to destroy the very heart of our village.
—Reverend A. Bronson, 1844

On a very hot, dry Sunday, two days before the Fourth of July in 1843 and a little before 4:00 p.m., two boys, who some say should have either been in church on this Sabbath day or under parental supervision, were celebrating the coming anniversary of the founding of the United States a little early with a toy cannon in an open space in the rear of a large three-story warehouse. Little did they know that their revelry would ignite the first great conflagration to befall their community and would scorch almost twenty acres of a new and vibrant downtown business district, destroy a legendary artifact made world famous in a poem by Henry Wadsworth Longfellow and leave nothing but ashes and a future city motto in its wake.

In 1843, Fall River was a still a town, not yet incorporated as a city, with some seven thousand inhabitants.[1] Fall River was founded thirty years prior, in 1803, when a petition was granted for it to separate from the town of Freetown to the north. The year when Fall River came into being saw the United States doubling in size through the Louisiana Purchase, Ohio becoming the seventeenth state, the U.S. Supreme Court first ruling a law unconstitutional (*Marbury v. Madison*), the first public library opening in Connecticut and Britain declaring war on France after Napoleon Bonaparte's continual interfering in Italy and Switzerland.

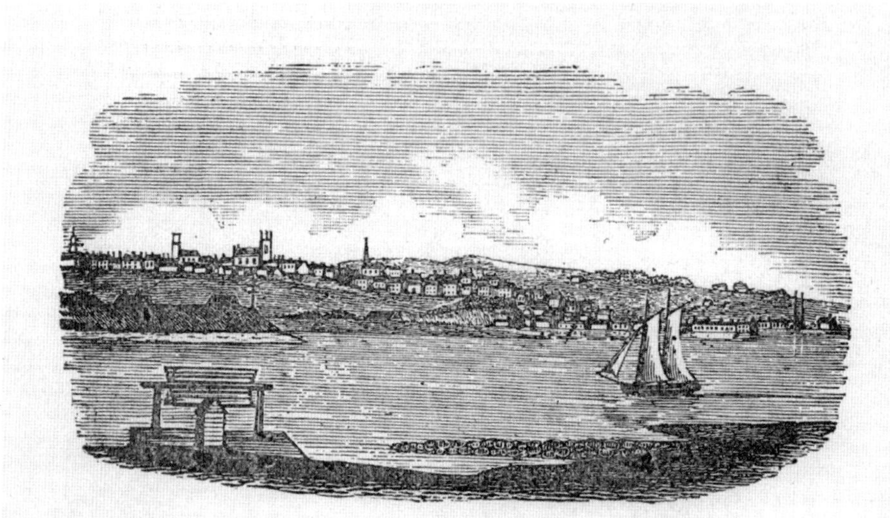

This view of Fall River, Massachusetts, is reproduced from *Barber's History of Massachusetts*, printed in 1838, the original sketch having been made about 1835, some eight years before the Great Fire of 1943. *From Lynch,* History of the Fall River Fire Department, *1896.*

By 1803, there were eighteen dwellings and one hundred residents in the town of Fall River,[2] and the first town meeting was held on April 4.[3] Yet in 1804, for some inexplicable reason, the name of the town was changed to Troy and remained so for thirty years while the village near the mouth of the two-and-a-half-mile-long Quequechan River was still referred to as Fall River.[4] In 1834, Troy once again became Fall River, mostly, they say, because the name had been fixed in people's minds for so long when it was part of Freetown. It can be argued that the village elders also realized that the "river" in Fall River (the Quequechan) was an important element in future industrial growth and that it would be a good idea to promote this fact to outside business concerns and those with a financial interest in investing in the ever-burgeoning commercial district. Some locals say, however, the change back to Fall River from Troy had a quite a bit to do with the sensational hanging death of the pregnant twenty-nine-year-old mill girl Sarah Cornell (who, coincidentally, was born in 1803) and trial of Methodist minister Ephraim Avery for her murder in 1833, a connection to the town they somehow hoped would be forgotten with the name change. It wasn't.

Before expounding on the details of the Great Fire of 1843, however, a bit more history is in order, if only to place this first disaster in context with just how far Fall River had come in such a short amount of time.

Fall River, as a community, transitioned from farm to factory in 1811, when the first cotton-spinning mill opened in what is now the Globe Village on the current site of Father Kelly Park. Colonel Joseph Durfee, a Revolutionary War veteran who took an active part in the Battle of White Plains, started this first mill. It operated as a cooperative—the cotton was imported from the South, distributed to the farm families for the intensive task of picking it, sent to the mill to be spun into yarn and returned back to the farmers to be woven into cloth. Finally, the cloth would again be given to the mill for finishing and made ready to sell. The first three cotton mills in Fall River were cooperatives. It was only in 1822 with the organization of the Pocasset Manufacturing Company—built on the second falls of the Quequechan River, the third mill at that location—by Samuel Rodman of New Bedford that the cooperative model ceased to be used.

In short order, two additional cloth mills opened: the Troy Cotton and Woolen Manufactory, located at the top of the falls, and the Fall River Manufactory, built on the third falls above tidewater. The Fall River Manufactory, the first cotton-spinning organization on the Quequechan River, was organized by David Anthony, who had the foresight to recognize the potential of opening a mill on the Quequechan River and using the natural power of its downhill flow to operate his machinery. There were eight waterfalls in the final half mile of the river's course. The flow of the Quequechan is said to be 121.6 cubic feet per second, or over 9,841,500,000 imperial gallons per year.[5] With this much power, it is easy to see why the river was so vital to these mill enterprises. Anthony's mill was 60 feet long by 40 feet wide and three stories tall with the lowest level made of granite and the upper floors of wood. The Troy Cotton and Woolen Manufactory was even larger at 108 feet by 37 feet and four stories in height.[6] This trend in building taller and longer mills would continue throughout the century.

By 1813, when these two mills opened, the up-and-coming town of Fall River boasted thirty homes, two hundred residents, three sawmills, four gristmills, a fulling mill, a blacksmithy and some small stores.[7]

Fall River proved to be an ideal location for cotton textile manufacturing, not only because its climate is quite similar to the English textile region of Lancashire, with higher relative humidity and more even and milder temperatures, but also because of its prime location on the south coast of Massachusetts on a waterway that empties to the sea. Transporting raw cotton into Fall River, instead of points farther north like Boston, meant that ships did not have to navigate the treacherous Cape Cod shoals. Other textile manufacturing cities like Lowell or Lawrence were located inland, on the Merrimack River. Because of lower transportation costs and faster delivery of coal shipments, the cost of manufacturing textiles in Fall River was 13 percent lower than in the inland Merrimack River centers of Lowell, Lawrence and Manchester, New Hampshire.[8] Additionally, shipping the cotton textiles to market in Providence, Philadelphia and New York was a relatively easy venture—a convenience for manufacturers, distributors and retailers.

Another event that would find its way to be part of the Great Fire of 1843 occurred in May 1831. In that year, housewife Mrs. William Cook—née Hannah Borden, the aunt of Andrew Borden (the father of Lizzie Borden)—was digging in the bank of the Quequechan River, near present-day Hartwell and Fifth Streets, for sand to use to scour her pots. The story goes that a section of the bank gave way, and Mrs. Cook was soon staring into the empty eye-sockets of a human skull. With the assistance of her husband and a family friend, clockmaker John Orswell, the skull and its attached skeleton, found in a sitting position, was unearthed.

One of the earliest accounts of the find appeared in 1837 in the *American Magazine of Useful and Entertaining Knowledge*. In it, writer John Stark, who personally examined the remains in the Troy Athenaeum, tells us that the bones were covered in a coarse dark bark that, when removed, revealed another layer of coarse cloth made of fine bark "about the texture of a Manilla [*sic*] coffee-bag." On the sternum was a "plate of brass, thirteen inches long, six broad at the upper end and five at the lower." The plate looked to have been cast "and is from one eighth to three thirty-seconds of an inch in thickness." Oval in shape, the breastplate was so corroded that the edges had become irregular and there was no way to know if anything had been engraved on it.

Below this breastplate was a belt made of "brass tubes, each four and a half inches in length, and three sixteenths of an inch in diameter, arranged longitudinally and close together; the length of the tube being the width of the belt." The tubes were made of a thin brass, "cast upon hollow reeds, and were fastened together by pieces of sinew." Stark surmised that the belt was used as protection for the lower parts of the body. Arrows were also found with the skeleton. Made of brass, they were flat, thin and triangular in shape, "with a round hole cut through near the base. The shaft was fastened to the head by inserting the latter in an opening at the end of the wood, and then tying it with a sinew through the round hole." He goes further to state, with some degree of certainty, that native tribes never practiced this mode of construction. While true, his opinion does not account for any sort of trading the Indians may have engaged in to secure these items. The arrows

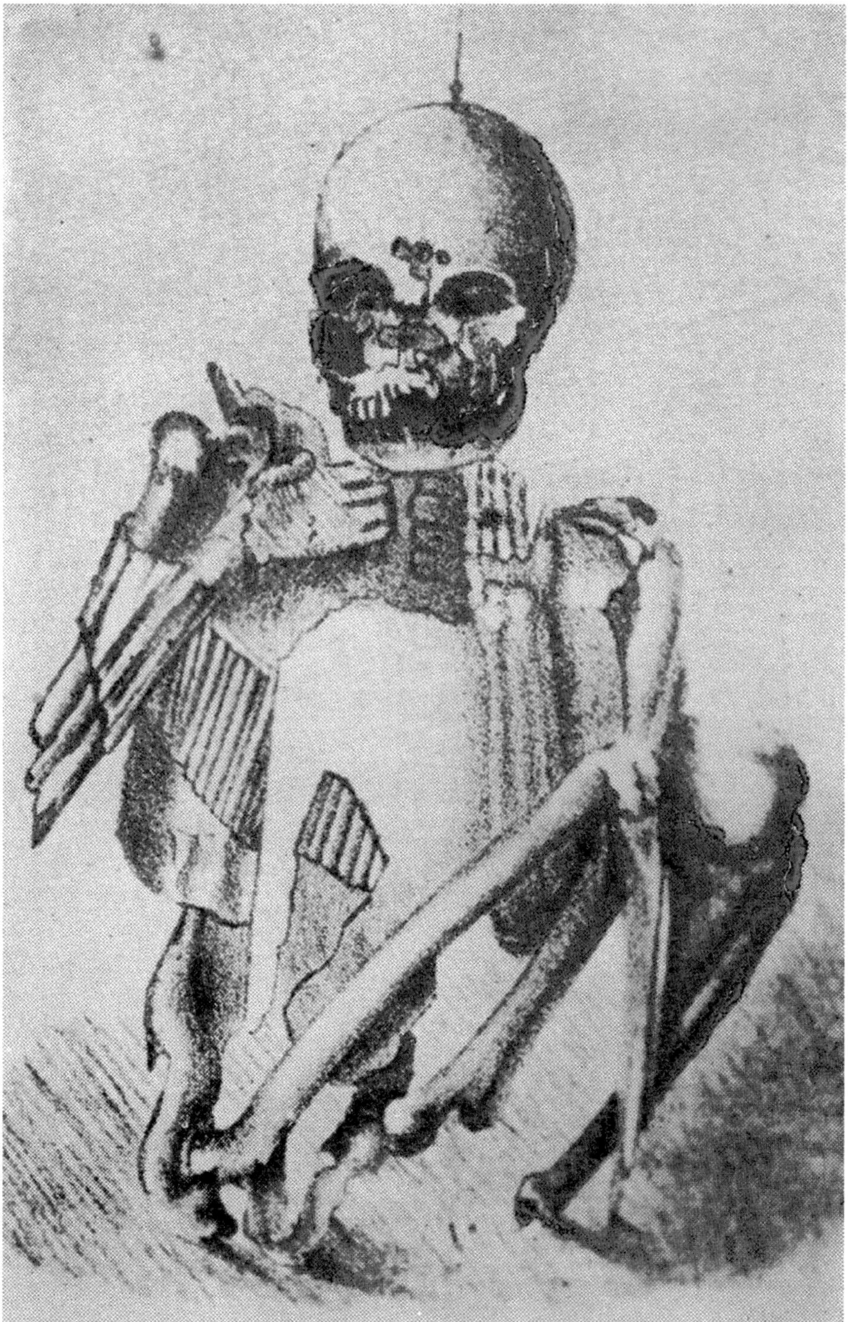

The "Skeleton in Armor," made famous by a poem by Henry Wadsworth Longfellow, was destroyed in the Great Fire of 1843. *Fall River Historical Society.*

were apparently in a quiver made of bark that fell to pieces when exposed to the air.

Stranger still, considerable flesh was still present on the hands, arms and back; under the belt; and for two inches above and below it. The skin had "the appearance of being tanned." In the body, apparently once a young man of approximately five and a half feet, the upper viscera of the chest "[was] probably entire." This preservation led Stark to hypothesize that either some embalming process had been carried out or this was the "accidental result of the salts of the brass [chest plate]" during oxidation.

Enter Henry Wadsworth Longfellow, who visited Fall River in 1840 to see his brother, the Reverend Samuel Longfellow, minister of the Unitarian Church. He brought with him as a companion Julia Ward, later to be Julia Ward Howe when she married. She wrote to her father of having seen the famous Skeleton in Armor. Poet Longfellow was so taken by the discovery that he penned his famous twenty-stanza ballad that romanticized the events and created a Viking tale that celebrates, in metaphor, the history of America.

We return now to our telling of the Great Fire of 1843, which stamped itself in the minds of the residents as such an important event that ever after

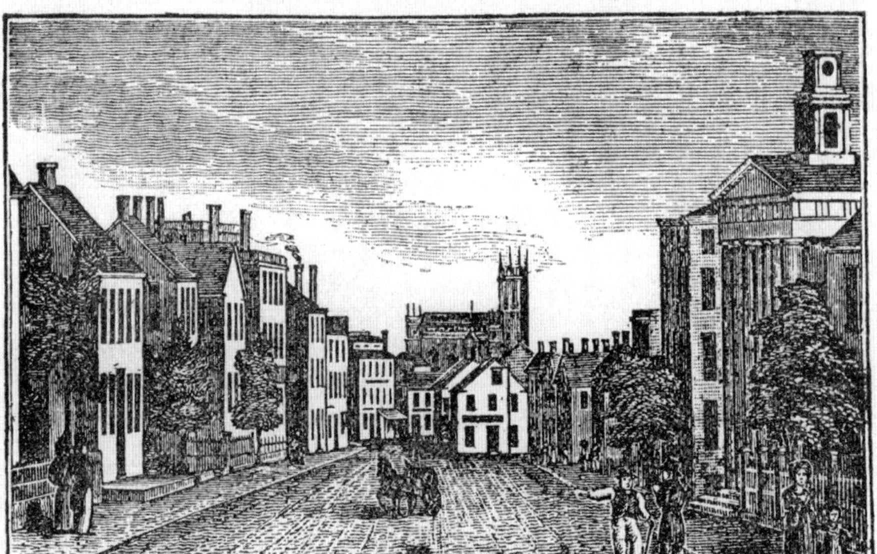

Main Street, Fall River, looking south from Pine Street, before the Great Fire of 1843. *From Lynch,* History of the Fall River Fire Department, *1896.*

this time period was marked by this generation as "before the fire" and "after the fire."[9]

At 4:00 p.m. on July 2, 1843, the thermometer stood at ninety degrees. Everything in the village was dry and parched "almost to a crisp" from a prolonged drought that had befallen the region.[10] There was a high wind, which chroniclers cited as something close to a gale, blowing in from the southwest. In another stroke of bad timing, the water in the stream (Quequechan River) had been shut off to allow repairs in some of the mills. Obviously, this was not the time to go mucking about with gunpowder.

In the rear yard of a warehouse occupied by Abner Westgate, on the corner of Borden and South Main Streets across from the Methodist Church, two boys, whose ages and names were mercifully never recorded for posterity, were celebrating the coming Independence Day by firing off a small toy cannon. The space where they were playing was covered with wood shavings, which were quickly kindled by the explosion. In five minutes' time, the flames were fifty feet high.[11]

The fire alarm was given at 4:00 p.m., which amounted to shouting and the ringing of bells, as it was a time before electricity or mechanical fire alarms. This was an age when "speaking trumpets" were employed by officers to amplify their voices over the noise and commotion of a fire scene and to direct resources to effectively fight a blaze. The Fall River Fire Department was formally established in 1832, eleven years previously, and was manned by an entirely volunteer company. Members were elected under the approval of fire wardens, and attendance at fires was mandatory, subject to fines of fifty cents for noncompliance, a large sum for 1843. Members received a small yearly compensation for their service, amounting to a cancellation of their poll tax.[12]

The first fire engine was bought by the Town of Fall River in 1818 and stationed on the south side of Central Street. It was a bucket-engine, which was designed as a means to spray water onto a fire with force and accuracy. However, with no process of drafting or sucking the water from a source, bucket brigades—a line of men who passed buckets of water from one to another—continually supplied the tub or cistern on top of the wagon. A simple piston pump—operated manually by pushing up and down long bars, called breaks or pumping arms, that ran parallel to the body of the engine—was used to force the water out of the tub through a nozzle and onto the fire. A full up-and-down motion of the pumping arm was called a stroke, and these engines operated at more than sixty strokes per minute. At this rate, a man could do this particular job for no more than a few minutes at a time.[13]

A carriage, with a large number of empty buckets, attended the bucket-engine, but many citizens had their own fire buckets, generally made of leather and painted with the name of the owner. The men were entrusted to bring their buckets with them when responding to an alarm.

The fire apparatus owned by the town in 1843 consisted of one suction hand engine; one bucket hand engine; one bucket carriage; one small hook-and-ladder truck; one rapid engine, stationed on Second Street (near where the fire started) with water supplied from a well; and a Hydraulion engine.

Manufacturing establishments, including the Fall River Manufactory and the Troy Mill, had in-house organizations known as Forcing Pump Companies, whose men were approved by the fire wardens. These "brigades" used hand hose reels and attended all fires that their hose, when attached to pumps at their factories, would reach. These volunteers also received compensation from the town and were likewise subject to fines for nonattendance at fires.[14]

The only means of fighting the 1843 fire, the "first general catastrophe in our local history," then, were hand engines and bucket brigades.

From the fire's source at Borden and South Main Streets, showers of sparks, cinders and firebrands carried by very heavy winds blowing nearly parallel with the street quickly ignited buildings on both sides of Main Street. Efforts to check the spread of the fast-moving fire were completely ineffective.

The town's Hydraulion engine was taken to the Quequechan River to draw water, but the flames moved so fast toward it that it was removed and "saved from the flames."[15]

According to Reverend A. Bronson in 1844:

> *The consuming element spread in every direction and with such amazing velocity, especially in the leeward, that no one could predict when or where this fire would cease! The goods of our merchants, and the furniture of our citizens, were scattered along the sidewalks from one end of the village to the other. And many, after removing their effects several times, had the agony to see them burned in the street at last. In vain did our Fire Wardens—sweltering under the combined heat of the sun and the fire—hurry from street to street to give their orders. In vain did our citizens labor and sweat to arrest the progress of the flames. All human effort seemed but mockery.*[16]

The fire extended north as far as Purchase and Franklin Streets, some seven blocks, and east as far as Rock Street, burning every building in its wake within this area. According to the newspaper that week, the "air filled with flakes of fire, and one building after another was in flames, till it seemed that if nothing short of a miracle could save the town from one general sweeping devastation."[17]

Six hours after the first alarm, about ten o'clock in the evening, a vessel arrived at the wharf from Bristol, Rhode Island, with a hand engine and fire company and assisted in saving houses on Purchase Street and prevented farther progress of the fire. The last house burned, incidentally, belonged to H. Battelle on Purchase Street and was the only structure north of Franklin to be destroyed.[18]

Almost all sources agree that if it had not been for the change in the direction of the wind, which reversed, causing the fire to blow back upon itself, the entirety of Fall River would have been in ruins.

Mrs. Seth Durfee described the events of that terrible day in a letter to her mother, Mrs. Edward Blanchard of South Weymouth, Massachusetts. Mrs. Durfee was attending an afternoon prayer service in Pocasset Hall when the fire alarm sounded in the middle of the sermon. She hurried to her home and found her husband sleeping. She cried to him, "What, lying asleep and all of Fall River burning up?" She and her husband assisted those in the path of the fire store their valuables. Luckily, their home escaped the conflagration. Describing the fire at its height, she wrote:

> *It was an awful spectacle to see and hear the raging, roaring and crackling of the flames, the rapidity with which they spread defying all exertion of the fire companies. The consternation of the people flying from one home to another beggars description...We do not know where the fire would have ended, had not the wind providentially changed just before night had blown the flames back upon the burnt district...Our people were entirely exhausted...I wish you could see Fall River the day after the fire. You could not tell where you were.*[19]

She related that Major Bradford Durfee, philanthropist and chief fire warden, had died suddenly one week after the fire. His death was attributed to overexertion suffered in fighting the inferno. His son, B.M.C. Durfee, in whose honor the future high school would be built and named, was but an infant at the time. "It was a sad and solemn day for Fall River," wrote Mrs. Seth Durfee.

Around midnight, Fall River's first conflagration was finally over, and almost two hundred buildings were in ashes. Nearly 20 percent of the population, 1,300 people, were homeless. Remarkably, there was not a single loss of life. In an effort to express their determination to not let this catastrophe stop them from succeeding as a community, the townspeople

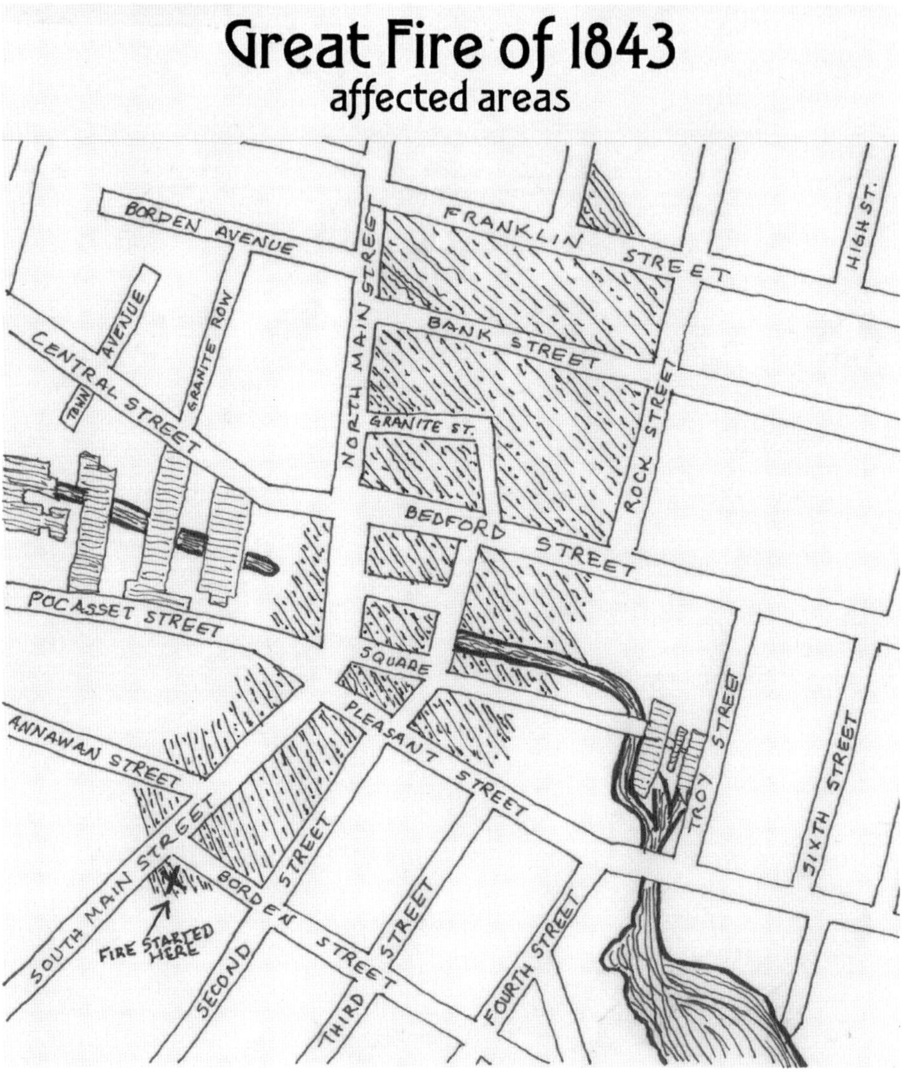

A map of the affected areas of the Great Fire of 1943 created from examination of the records of the event by the author.

created a banner and hung it in the heart of the ruins. The message was simple and to the point: "We'll Try."[20]

The final estimate for the size of the fire was that the burnt district encompassed twenty acres of the center of business operations. Among the buildings consumed, including the Troy Athenaeum, where the Skeleton in Armor was housed, were three churches and all public offices.

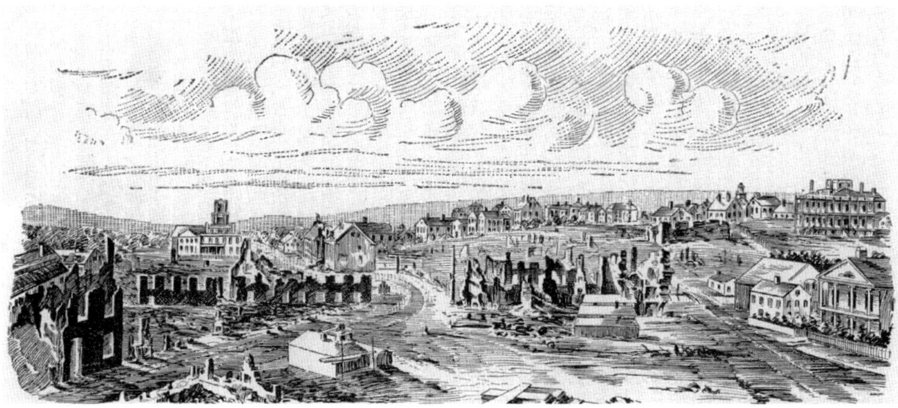

This lithograph by E. Ruggles, a textile artist, shows a view looking north from Borden Street of the burned district of the Great Fire of 1843. In the foreground are the ruins of the Bridge Mill. Above them, intact, is the Stone Church. On the right is the Nathaniel B. Borden mansion. In the middle are the ruins of the Pocasset House, a hotel. *From Lynch,* History of the Fall River Fire Department, *1896.*

> *Our post office and custom house are gone; and we have not a printing office, nor hotel, nor bank building, nor book store, nor market, nor bakery left. Nearly all our grocery and provision stores, including one wholesale establishment, with most of their contents; and all our dry goods, druggist, tailor, milliner, tin ware, and paint shops, with one cotton factory, running 3,000 spindles are gone.*[21]

A general mass appeal was sent out for aid that requested both dollars and donations of goods not only to help those who had lost everything in the blaze but also for rebuilding the town. "Prompt and generous donations of money, clothing and food were received in quantities sufficient to prevent immediate suffering." In fact, money poured in from fellow Americans. In all, $50,934 was raised from communities like Boston, New Bedford and New York City.[22]

A fire committee was appointed to determine a correct list of those who suffered and the amount of property destroyed. It published its report the following year. In it was listed every property that was damaged or destroyed and the dollar value of the losses. The tally at the conclusion of the report was sobering:

Whole amount of loss on buildings: $264,470
Whole amount of loss on other property: $262,015

Total loss on both: $526,485
Whole amount of insurance: $175,475
Excess of loss over insurance: $351,010[23]

The total excess of loss over insurance is a huge figure for the year 1843. Converting this amount to 2014 dollars would mean that the Great Fire of 1843 resulted in an approximate $140 billion loss in buildings, personal property and merchandise.[24]

Fall River rebounded and eventually thrived from the greatest disaster of the era. In short order, rebuilding commenced, and within a year, the town proudly boasted that its reconstructed community was even better than what had been there before. Said the Reverend A. Bronson in *An Address on the Anniversary of the Fire, Delivered to the Pearl St. Christian Chapel, July 2nd 1844*:

> *Our streets have been widened and straitened; and noble structures and elegant massive blocks, now appear where desolation so recently held empire. A splendid Market and Town Hall, with two more magnificent edifices, which are to occupy the places of the Pocasset House, and the Pocasset Factory, will cover completely the whole territory of our ruins. Besides, five houses of public worship have been erected in a superior style. It is true of these, that the external beauty and glory of these latter houses greatly exceed the glory and beauty of the former houses.*
>
> *But this is not all. In other parts of our village, as well as in that portion which was scathed by the fire, many, very many houses and stores have been built, and others are now in the progress of erection. So that visitants from the neighboring towns can hardly recognize Fall River, as an old acquaintance; and some of our citizens have been heard to say, that they find themselves lost, at times, in their own streets. Indeed, our ruins have been restored, and much more than restored, as if by enchantment!—The heart must be cold as ice that does not glow with gratitude to God on this anniversary of the Great Fire.*[25]

Eleven years later, in 1854, Fall River was finally incorporated as a city—the eleventh city in the commonwealth of Massachusetts. The city's first mayor was James Buffington, a former fire warden and chief engineer of the Fire Department. Fall River had grown to have a population of more than twelve thousand, a valuation of $8,939,215 and an annual tax of $56,000.[26]

When the new city designed its seal, its citizens wished to incorporate elements in the graphic for which the city is known: the Taunton River,

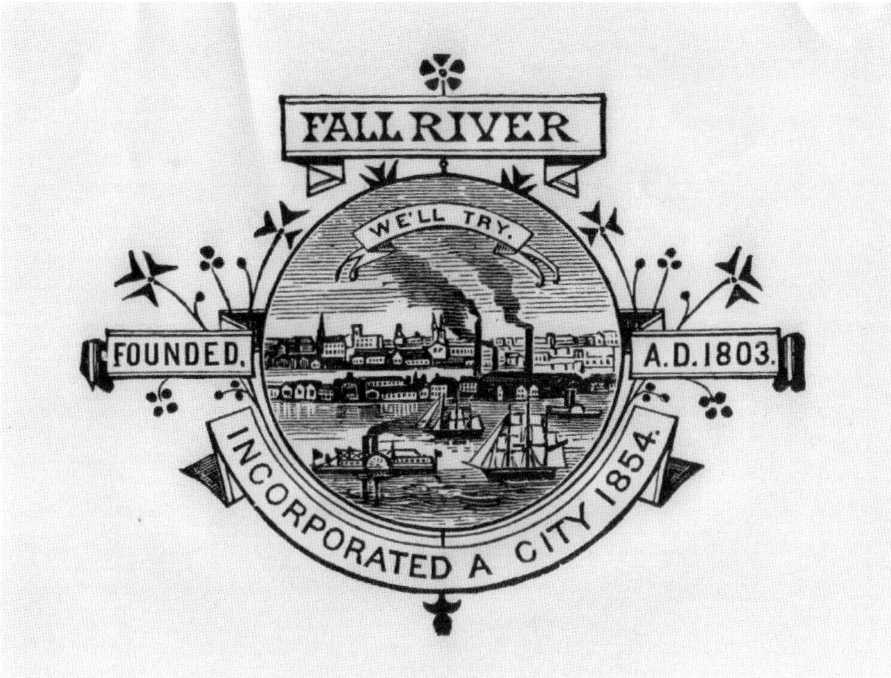

Seal of the city of Fall River, Massachusetts. The motto, "We'll Try," was born following the Great Fire of 1843 to describe the city's determined effort to rebuild following this destruction. *Fall River Public Library.*

church steeples, textile mills, the year the town was founded, the date of incorporation and a bustling harbor. Prominently displayed in the image are also the words "We'll Try," placed there to remind succeeding generations that it is, indeed, possible to overcome a citywide disaster and even prosper in its wake—as long as there is a will to do so.

2

THE GRANITE MILL FIRE OF 1874

It was a calamity unparalleled in the annals of the city, and transformed many a happy dwelling into a place of mourning and suffering.
—History of the Granite Mill Fire *(1874)*

The Granite Mill was the twenty-third textile enterprise incorporated in Fall River and the first established during the Civil War period. Named for the native construction material quarried on site, the company was founded by Charles O. Shove, William Mason and Southard H. Miller as principal investors in 1863 with a capital of $225,000, raised to $400,000 by 1871.

Mill No. 1 was constructed in 1865 along Twelfth Street, between Pleasant and Bedford Streets, and was 328 feet long by 70 feet wide and five and a half stories tall. It boasted a central, Italianate-styled tower, which contained the building's only entrance and stairway; a steep-pitched barn roof; and an attic, having a window that opened to the roof. A single fire escape was located on the east side of the mill and reached to only the fifth floor. When it opened, Mill No. 1 had 33,866 spindles and 860 looms. The mill was powered by a 650-horsepower double Corliss engine, fed by twenty-four cylinder boilers, with water for steam purposes drawn through a canal from the Quequechan River.[27] The "mill was built in accordance with the best plans then in use, and was regarded as a model of its kind."[28]

Granite Mill No. 2 was built in 1871 at the corner of Bedford and Robeson Streets and was nearly identical to Mill No. 1 but with an additional fifty feet in length. Mill No. 2 contained its own boiler house

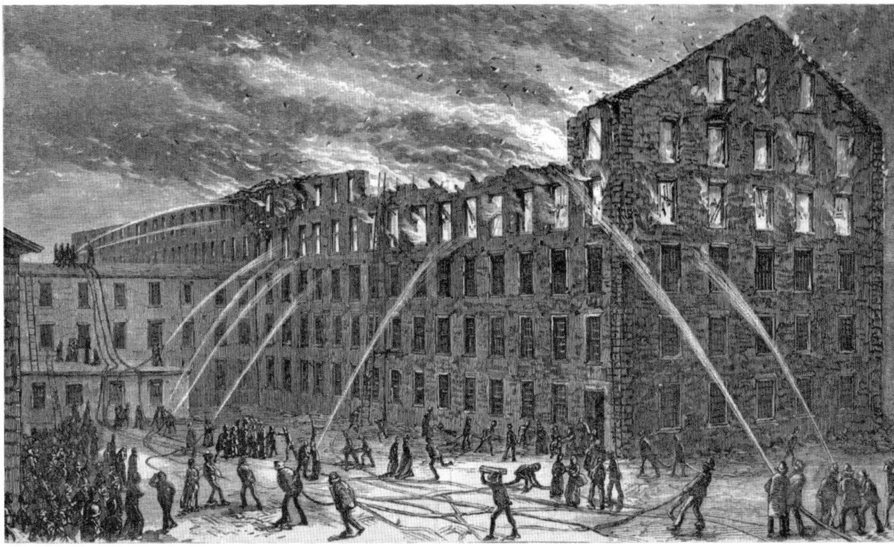

"The Ruins of the Granite Mill." *From* Harper's Weekly, *October 10, 1874; sketch from photograph by E.F. Gay, Fall River.*

and was powered by a 750-horsepower double Corliss engine, with 44,664 spindles and 1,008 looms.[29]

In Mill No. 1, the first and second floors were occupied by weaving apparatus, "containing about five hundred looms each,"[30] and a machine shop. The third floor was devoted to carding and roving.[31] The fourth and fifth floors were for spinning, with the fifth floor containing sixty-four spinning mules.[32] The attic, or half floor above the fifth, housed apparatus for drawing in, spooling, slashing and warping and also contained one pair of mules.[33] "On this floor was a large number of persons employed compared to the others."[34]

In his book *Natural and Manmade Textile Fibers*, George E. Linton details how cotton is transformed from cotton to fine thread. The creation of cotton textiles begins with the planting, harvesting and ginning of cotton, when the cotton lint and seed is separated. Ginning is essentially a cleaning process; raw cotton moves through dryers to reduce moisture content and then passes through the gin to remove foreign matter. Only one-third of the harvested cotton crop is cotton fiber, Linton says, the remaining two-thirds being waste matter, such as burs, dirt, stems and leaves. The first ginning process takes place at the plantation cotton gin or at the "community gin." The resulting lint is brought to the major cotton cities of the South, where it undergoes a second ginning procedure and is compressed into cotton bales, each about five hundred pounds.

Most of Fall River's textile mills received the cotton bales, processed the cotton into yarn and wove the yarn into gray goods fabric, called mull. However, some mills in the city specialized in manufacturing thread (like Kerr Mills), some bleached the gray goods into white fabric (as did the Fall River Bleachery) and some mills concentrated their manufacturing operations in printing the white cloth (like the American Printing Company). Occasionally, all operations, from raw cotton to printed textiles, were conducted on the same site, as was true of the Chace Mill on Rodman Street. After 1895, that mill added bleaching, dying and finishing to its operations to complete the cycle.[35]

Granite Mill No. 1 employed four hundred operatives, many of whom were women and children. "Child labor laws had not yet freed them from a life susceptible to exploitation, disease, and early death."[36] In 1842, Massachusetts passed a law limiting children's workdays to ten hours, but this law was not consistently enforced. Federal law regarding child labor was not passed until 1938 with the Fair Labor Standards Act. Interestingly, it was the photographs of Lewis Hine, who traveled the country for the National Child Labor Committee and recorded instances of child labor in mills, that helped awaken public and government sentiment against the use of children's labor in industry. Hine visited Fall River in 1911 and again in 1916, photographing over two hundred images of children at work in the city. His pictures were instrumental in reforming child labor laws in the United States.

David Lawler recounts his years of working in the mills of Fall River in his *Life and Struggle of an Irish Boy in America*, beginning with his first job when he was but six years of age. His family emigrated from Ireland, with his father traveling ahead of the family to see if opportunity existed for him as a mechanical engineer. He landed a good job in Boston, but "his friends advised him to go to Fall River to work for there would be work for the children. And the children worked in the cotton mills. All of us, seven boys and three girls, did. Two of my brothers died in infancy, so they escaped mill life."

Lawler's first job was in the mule room of the Granite Mill No. 2 in August 1872. He was paid thirty-two cents a day for his labors.

> *At the time, the mill hands began work at six o'clock in the morning and worked until six-thirty at night. I helped to put bands on the new mules and then became a tuber and a doffer. After being there three or four years instead of getting thirty-two cents a day I got twenty-one cents for doing the*

> *same job—everyone was paid accordingly. After I had worked there for ten years I got forty-eight cents a day. I could not go any further unless I became a mule spinner and that would be ten dollars and a half a week. While I was clever, being champion doffer and tuber in all contests, the gap was too great. I would have to wait five or six years more to get the necessary strength, so my brother got me a job in the Devol mills, where I got a dollar a day making ginghams.*[37]

The job of a mule spinner was arduous and grueling, so adult men almost always performed this duty until the 1880s, when even more women began to join the workforce. By 1900, in some mills in New England, all of the spinners were women.[38]

> *A mule spinner looked after two mules. Each mule controlled about one thousand spindles. The mule would travel out on steel tracks, which resembled a small way railroad tracks, a distance of about five feet. The spindles turning about ten thousand times a minute. Thus the thread was made from the drawing of the rolls and the twisting of the spindles. At the end of the run the mule would reverse and wind the thread on the cop. The spinner's work was to piece the broken threads. This kept him on the jump all day when the cotton was good, but when the mill treasurer bought inferior cotton the work was very hard...We boys were assistants to the spinners; we kept the ropen* [roving] *in back of the mule. It was from this the thread was spun. We took off the full cops and put on the tubes for another beginning. We took away the full boxes of yarn and weighed them, for each man was paid by piece work.*[39]

The population of Fall River in 1874, the year of the tragedy, was over 45,000, and it would only grow from there, more than doubling by the turn of the century to 104,863.[40] It is estimated that in 1875 nearly 32 percent of the adult population, both male and female, was employed in the cotton mills, not counting children, who are known to have been a substantial part of the workforce.[41] Between 1845 and 1875, forty-two cotton mills with 1,268,000 spindles were in operation—twice the number operating in Lowell. In 1871 and 1872 alone, fifteen new cotton textile manufacturing corporations were established in Fall River. The Granite Mill No. 1 was regarded as a model of architecture and design, and in some respects, it undeniably was.

Early Saturday morning, September 19, 1874, at ten minutes before 7:00 a.m., a fire caused by a friction spark that kindled some cotton broke out on the north end of the fourth-floor mule room at the pair of mules No. 15.[42] Seventeen-year-old Dennis Leary, a spinner, had just picked the roller, oiled No. 15's mulehead and started the machinery when the fire broke out. The blaze appeared, at first, to be small and extinguishable, so he attempted to put it out with an available pail of water, one of the one hundred that were alleged to have been located strategically in the building and were to be used for just such an event. There had been fires before in the muleheads, and usually a bucket of water was all that was necessary to douse the flames.

However, one bucket of water was not enough. When Leary went for another, he found that fellow operatives had gathered to observe the fire, blocking the work aisle between the mules, and he was unable to pass by. During this momentary pause, the threads connected to the mule caught fire almost instantaneously and conveyed the blaze to all the mules nearby, "causing the flames to form a complete and impassible wall of fire from one side of the room to the other." The flames quickly swept up fifteen feet to the floor above. "It would be almost impossible to express the swiftness with which the barrier of fire spread itself, and the vigor with which it withstood the utmost endeavors of man and mechanical appliances."[43]

In a second attempt to extinguish the fire, Leary "grasped a coil of hose that was attached to his mule alley, opened the stop-cock and valve connected therewith, and called upon the men to go down and start the donkey engine," which would have provided a flow of water. By this time the fire was growing in momentum and created such a critical condition, that Leary passed the hose off to another man and rushed to the valve near his mule and opened it, expecting water to be forthcoming. Unfortunately, none came. Leary took the hose again and sent the man to order the water turned on, maintaining his position there until his empty hose caught fire. "He then ran to the door, but before he reached it he was compelled to drop on his hands and knees and crawl through the smoke that was now pouring out of the room in great volumes."

Sixteen spinners had been in the room when the fire first broke out. Nine men had run away at the first hint of smoke and seven others stayed to fight the fire with pails of water. Their efforts not being successful, they, too, were forced to flee for their lives.

By this time of the morning, all of the machinery was in operation on the fifth floor and attic, so employees in these upper stories did not hear any of the fire commotion until the flames had burst into the rooms in which they worked. By this time, they were completely trapped and "doomed to a miserable death."

Because the fire began on the north side, the location of the tower stairway, and swept along to the south side of the building, escape down the stairway was impossible. The women and children were stranded and cut off by the flames and smoke. The single fire escape was of no use, as it extended to only the fifth floor, and escape from that level soon became impossible, as well.

> *Half-maddened women operatives rushed hither and thither, shrieking wildly for aid, and imploring the men to save them. Some were praying, some loudly lamenting, and some were so paralyzed with fear that they stood as in a trance, seemingly oblivious to the fate which menaced them; while others, stood stock still and wrung their hands in the agony of despair.*

The second overseer organized the women in an attempt to get them down the tower stairway, but he found it impassable due to mounting smoke and flames. He backed up and, closing the door to keep the room as free from smoke as possible, rushed the women to the elevator. This means of escape was also prevented because the engine had been turned off, and the elevator would not work. The mill hands were now driven, as their final refuge, to the windows at the southern end of the building.

> *The second overseer had obtained a coil of rope which he threw out of one of the windows, and then tried to persuade the girls to go down that way; but, with a few exceptions, they were afraid to risk it, and preferred to remain at the windows until the biting flames and fast-accumulating smoke compelled them to adopt the more dreadful measure of leaping to the ground.*

An amazing and daring escape was attempted by a slasher tender and his companion, both employed in the attic on the north end of the building, when they became separated from the rest of the workers by the wall of flame. They found a stepladder and a length of rope and after one made his way to the roof through the skylight, he turned to his assist his friend to join him. The man tried many times to jump from the top of the ladder to the roof, but was unable to do so.

> *The ladder fell from under him, and he tumbled back into the burning room, never to rise again. The slasher tender then crept along the edge of the roof—which was an extremely hazardous feat, as it was slanted*

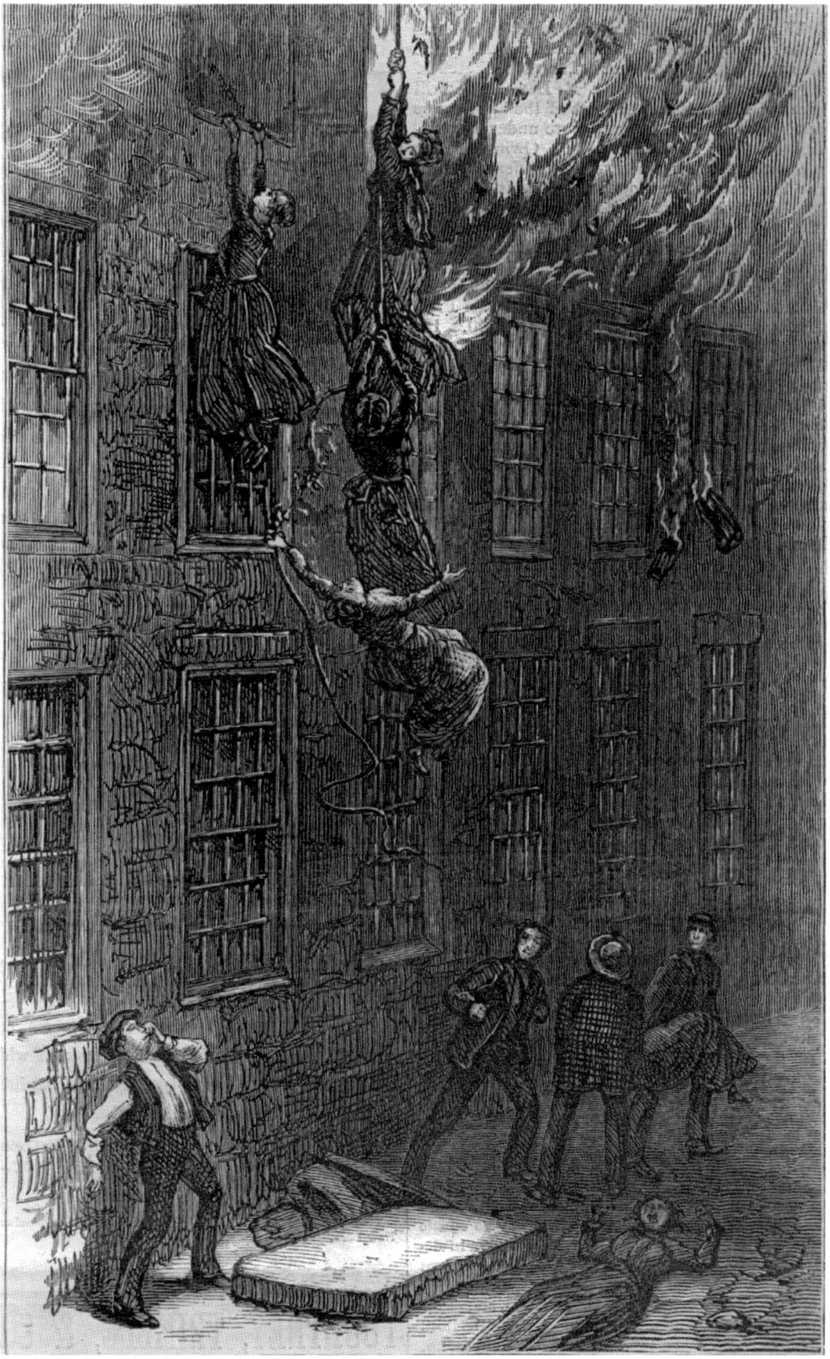

"The Panic—Escaping Through the Windows." *From* Harper's Weekly, *October 10, 1874; sketch from photograph by E.F. Gay, Fall River.*

> *precipitously and the foothold was very precarious—until he reached the verge of the northeastern corner of the building. He then fastened the rope to a piece of exhaust pipe and threw it over the side of the mill. But the rope was all knotted and twisted, and fell only about ten or twelve feet. Then he swung himself over, and slid down as far as the rope would permit, and dangled there, in mid-air, before the eyes of the anxious and excited spectators, while the flames and smoke were rushing towards him, threatening momentarily to enwrap him in their deadly embrace, or compel him to drop off and be dashed to pieces on the ground below. But he was a brave, cool, man, and made a gallant fight for his life. Managing to unloosen the coil with his feet, he slid down a couple stories further, until he met a similar impediment, which he overcame in the same manner, and then came down hand over hand until within a few feet of the ground, when he dropped into a dozen outstretched arms that were waiting to receive him.*

Nine-year-old David Lawlor remembered that day well. He was working in Granite Mill No. 2 at the time of the fire and saw a "puff of smoke coming out of the window and a man descending the fire-escape." One of the boys who had jumped from the sixth floor was a companion of David's who had been sent over to work in Mill No. 1 a few days before the fire. "He jumped, ran home, told his mother the mill was on fire, and dropped dead. Few escaped from that room."[44]

Granite Mill superintendent James E. McCreary testified at the inquest that when he came through the room where the fire started at twenty minutes to seven, everything was as it should be.[45] He exited the building and, while standing on the corner of North and Twelfth Streets, saw "fire and smoke issuing out of six windows on the north end, west side." His sworn statement regarding his activities that morning, who and what he saw as he ran about the mill, and his self-described heroic actions were all called into question when compared to the testimony of the others.

McCreary says that upon seeing the fire from the corner, he rushed into Mill No. 1 and "struck the alarm." Then he ordered a man to tell the engineer to turn off the gas, grinding all the machinery to a standstill. He next ran up the stairs in an effort to warn those on the top floors. He claims that when he reached the fifth floor, he pleaded with the workers to evacuate, "telling them that if they did not go quickly they would not be able to at all, soon." He asserts that he reached the attic and went in two-thirds of the way to the north, yelling at the top of his lungs, and then yelled the same a similar distance southward. He said there was no response, saw and heard no one and assumed

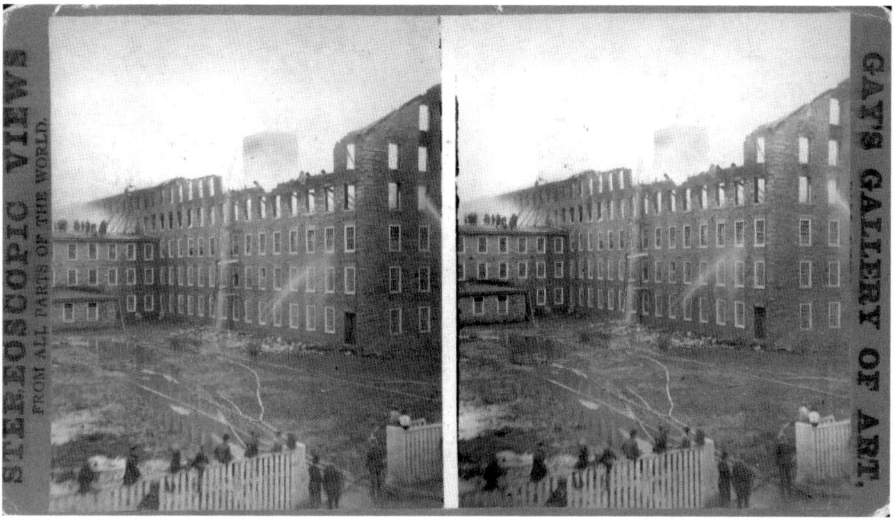

A stereoscopic card for the Granite Mill Fire of 1874, Gay's Gallery of Art in Fall River. *Private collection.*

that everyone had escaped. The smoke was so dense that he could not breathe, so McCreary tied a piece of cloth over his mouth and descended. He claims all this happened within six minutes of seeing the first fire.

On the way down, he met Mrs. Caveney on the fourth floor, whose husband worked in the spooling room. He told her she could not go upstairs and "turned her back. There was no one in the fourth story at that time."

McCreary testified that when he was in the attic, he observed that there was nothing to prevent the workers from coming down the stairway as he did. He went to the carding room and found a woman there who was looking for her shoes. He told her that he would buy her a pair to "clear out." He said he found workers hastening back into the lower rooms to retrieve their clothing and he "drove them out, stationing men at the doors to keep them out." He said he knew of no one who had gotten away by means of the fire escape.

McCreary also spoke of the water supply in the mill and its use in fighting fires, saying there was a pipe no more than fifteen feet from where the fire started that "only needed pulling down to let the water on." The mill had been designed, he said, with five streams of water in each story—two from tanks, which would have taken about one minute to get working. "There was nothing needed but just to open the valve."

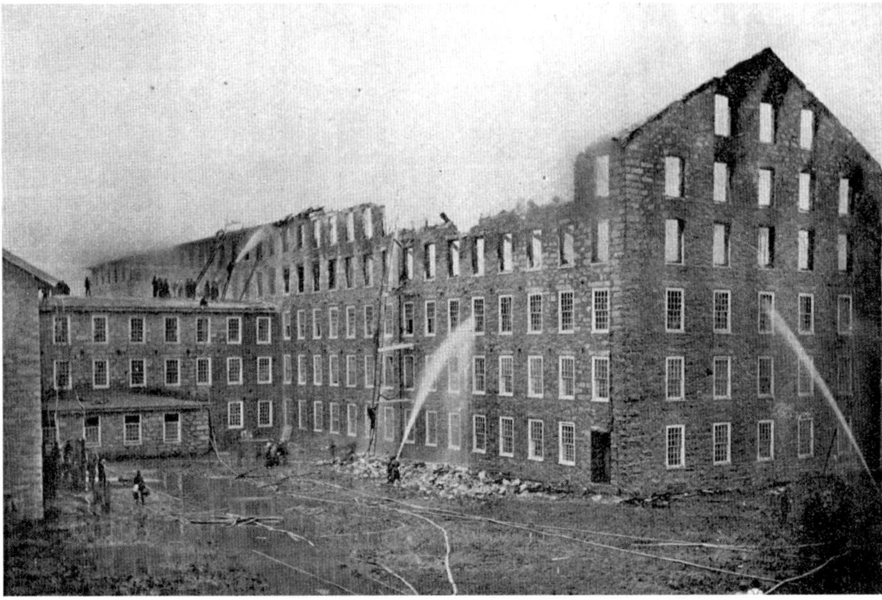

Granite Mill No. 1 during the fire of 1874. *Fall River Historical Society.*

In a chilling close to his testimony, McCreary added, almost as an afterthought, "Supposing there was no one in the mill, at the suggestion of one of the directors [I] set men to work bringing out the cloth."

The first alarm of fire was struck from Box 72, located on the corner of Sixth and Pleasant Streets, at 6:55 a.m. Almost at the same time, an alarm was also struck from Box 74, "causing some trouble on the 'tappers' in the engine-houses, and a consequent delay in locating the fire." The hook-and-ladder company was the first to arrive, having received the first round of the alarm correctly. Sadly, the firefighters were of no use. "The longest ladder in the department failed to reach anywhere near the attic windows, at which could be seen the poor creatures struggling with each other in vain efforts to escape their impending doom."[46]

The remaining members of the fire department were on the scene as soon as they had sorted out the confusion of the alarm that had struck four different boxes at the same time.[47]

Quick-thinking bystanders brought their bedding from their homes. Mattresses were piled up and blankets held by firemen and residents to help break the fall of those jumping. Horrifyingly, "some who jumped became impaled on the ladders below."[48]

Most of those who jumped landed wide of their mark, and some, it seemed to those watching, did not even look to see where they were leaping, but threw themselves out of the windows in a frenzy of fear.

Eleven-year-old Maggie Sullivan worked in the attic. Because it was Saturday, she had taken her younger sister Katie, aged eight, to help her, as was common. When Maggie realized the dire situation, she asked her sister if she wished to stay and be burned or dropped out of the window.

> *"Throw me out of the window," said little Katie, firmly. "I don't want to be burned." So Maggie took her sister in her arms and leaned out of the window to drop her on the ground, 65 feet below. As she was about to let her go, she observed a woman going down from another story on a rope, and being afraid that she might knock her off, waited, holding her patiently, until the woman was safe. Then she dropped her sister down. When the crowd below had removed her, she climbed to the window sill and jumped down herself, without the slightest hesitation. Surely, the good angel of the two wonderful children bore them up carefully. The smaller one bruised her breastbone somewhat, and the younger broke her ankle, and received several bruises. They both got well.*[49]

The three Murphy sisters, Bridget, Kate and Margaret, were not so lucky—all of them lost their lives in that same wild leap. David Lawlor remembers that it had been Mrs. Murphy's custom to bring her three daughters breakfast every morning. After the fire, and for some time, "she would get the breakfast, fill the pails, walk down to the ruined mill and cry hysterically until some friends would take her home again."[50]

A young boy climbed down a rope from the attic as far as the fourth story, when the friction burns in his palms, caused by his rapid descent, compelled him to let go. "He fell to the ground, where he was taken up with scorched limbs and a shattered arm." A small girl threw herself from a window and, to the horror of the onlookers, struck a ladder and hung there until she was taken down by some of the firemen. She was dead. A young man who had two sisters employed in Granite Mill No. 1 ran to the scene of the disaster and saw them at an attic window. He called to them to jump into his arms and one of them did so. Miraculously, he caught her as she leapt, which brought the brother and sister, uninjured, to the ground. "The other sister was afraid to leap, and perished in the flames." Another young child, upon seeing her mother in the crowd below, stretched out her arms "and called on her parent to save her—and then took the dreadful leap which landed her in eternity." A spectator who was

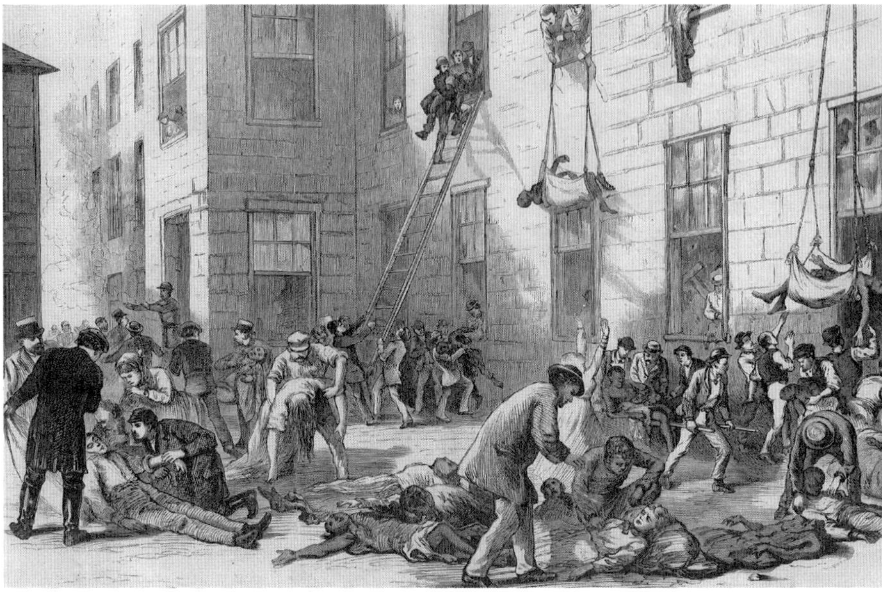

"Finding the Bodies of the Victims." *From* Frank Leslie's Illustrated Newspaper, *October 1874; sketched by E.R. Morse.*

holding a mattress "to catch the flying forms, was struck by the falling body of a woman and was severely injured, while the hapless female was instantly killed."[51]

The fire was finally extinguished at 3:00 p.m. It was then that the sad task of removing the dead commenced.

In all, twenty-three perished in the Granite Mill fire, a blaze reportedly visible from ten mills in the city. More than thirty others workers suffered serious and lifelong injuries, with many disfigured.[52] Those who jumped from the windows died from spinal injuries, crushed skulls, broken necks and shattered bones. The dead were mainly women between the ages of twelve and twenty-one and children between the ages of ten and twelve. The youngest employees of the Granite Mill worked on the highest floors.

The bodies of the dead were taken to the firehouse, to nearby mills or to churches to await families' identification. The city began a week of mourning. Funerals for the dead brought out hundreds of residents who silently bore witness to the tragic consequences of the fire. The funeral procession for the three Murphy girls was particularly solemn. Even thirty-seven years later, the *Fall River Herald News* remarked, "And there are those who can still recall the three hearses and three trains of bearers passing through the street on the mournful Sunday that followed."

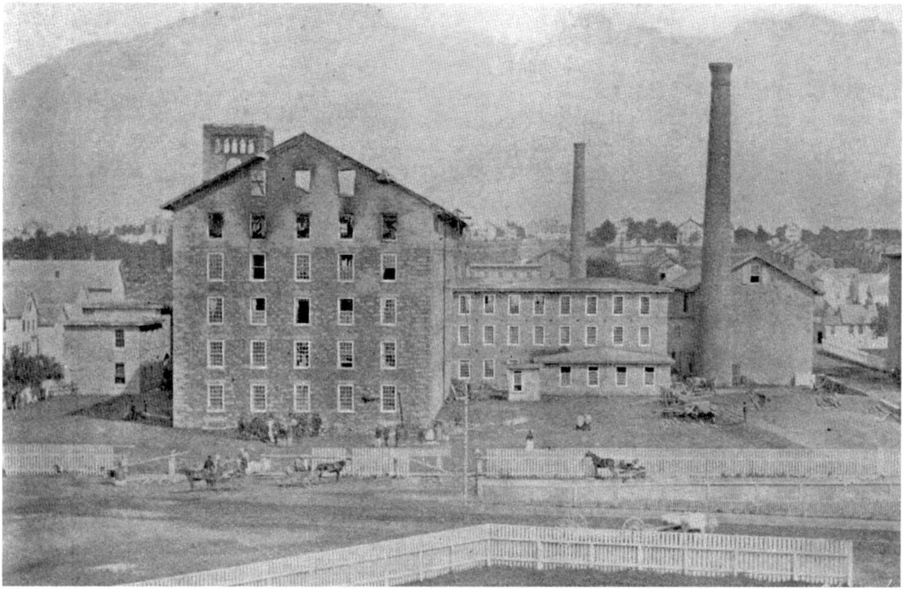

Granite Mill No. 1 following the fire of 1874. *Fall River Historical Society.*

In the wake of the deadly fire, it was clear that battle lines had been drawn. The workers blamed the policies of the mill managers, saying that the overseers denied the spinners time to clean and oil the machines. In addition, the scarcity of a water supply, inappropriate number of water buckets and the lack of escape routes all worked against the mill hands and their efforts for survival. The mill owners blamed the operatives for not having "the presence of mind to warn those on the 6th floor of the fire," for fleeing "in fear and haste" and for "not helping women and children to safety." The fire department also came under harsh criticism for their inability to rescue the workers and having apparatus that was not long enough to reach the upper levels of the mill.[53]

A jury of six was impaneled to conduct a coroner's inquest on September 21, and it met until October 3, with a ruling reached on October 5. The jurors interviewed forty-nine witnesses and took voluminous testimony. Their verdict exonerated the mill owners and the fire department, even as they criticized them for their insufficient handling of the disaster. Granite Mill No. 1, they concluded, "was built according to the best plans then in use and with what safety equipment was then in vogue. Fire officials, who prior to and after the fire, were critical of their own meagre [*sic*] equipment, reportedly used it as best they could."[54]

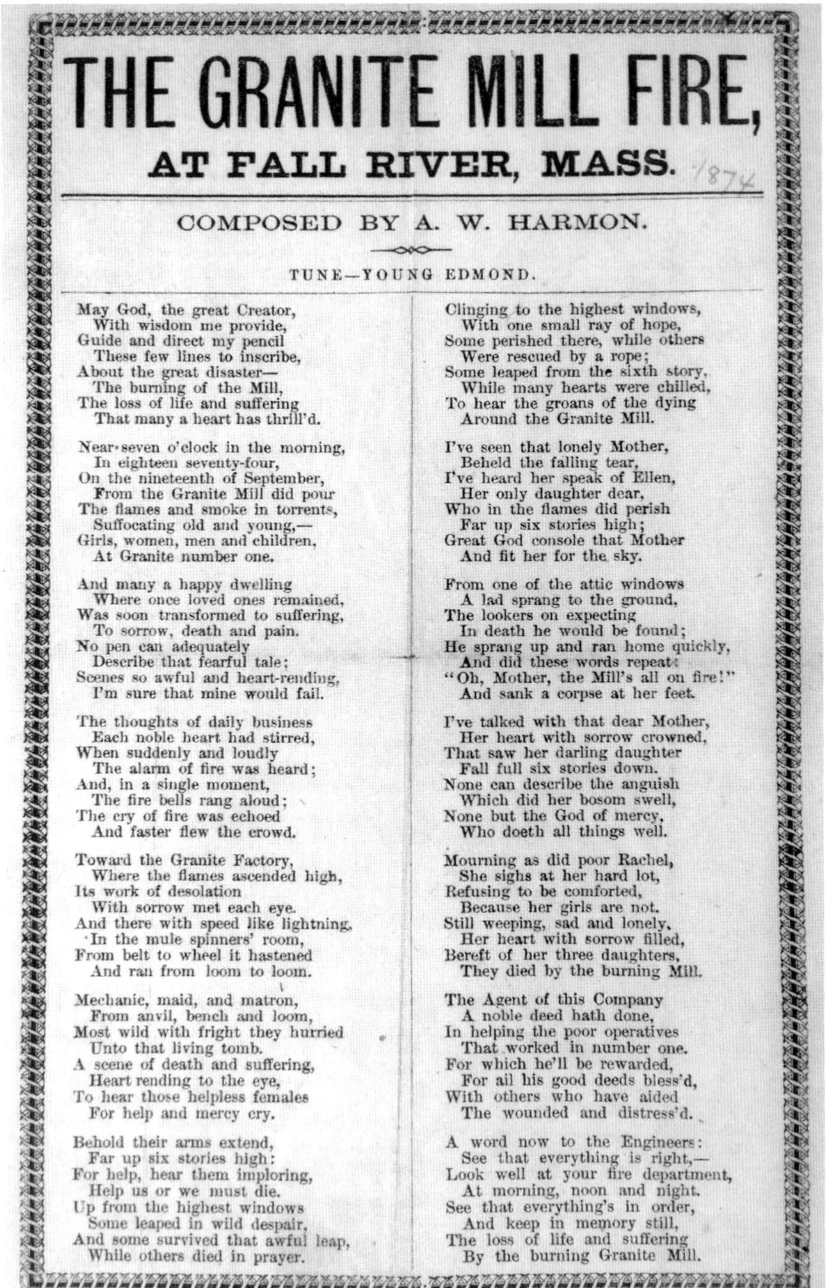

THE GRANITE MILL FIRE,

AT FALL RIVER, MASS.

1874

COMPOSED BY A. W. HARMON.

TUNE—YOUNG EDMOND.

May God, the great Creator,
With wisdom me provide,
Guide and direct my pencil
These few lines to inscribe,
About the great disaster—
The burning of the Mill,
The loss of life and suffering
That many a heart has thrill'd.

Near seven o'clock in the morning,
In eighteen seventy-four,
On the nineteenth of September,
From the Granite Mill did pour
The flames and smoke in torrents,
Suffocating old and young,—
Girls, women, men and children,
At Granite number one.

And many a happy dwelling
Where once loved ones remained,
Was soon transformed to suffering,
To sorrow, death and pain.
No pen can adequately
Describe that fearful tale;
Scenes so awful and heart-rending,
I'm sure that mine would fail.

The thoughts of daily business
Each noble heart had stirred,
When suddenly and loudly
The alarm of fire was heard;
And, in a single moment,
The fire bells rang aloud;
The cry of fire was echoed
And faster flew the crowd.

Toward the Granite Factory,
Where the flames ascended high,
Its work of desolation
With sorrow met each eye.
And there with speed like lightning,
In the mule spinners' room,
From belt to wheel it hastened
And ran from loom to loom.

Mechanic, maid, and matron,
From anvil, bench and loom,
Most wild with fright they hurried
Unto that living tomb.
A scene of death and suffering,
Heart rending to the eye,
To hear those helpless females
For help and mercy cry.

Behold their arms extend,
Far up six stories high:
For help, hear them imploring,
Help us or we must die.
Up from the highest windows
Some leaped in wild despair,
And some survived that awful leap,
While others died in prayer.

Clinging to the highest windows,
With one small ray of hope,
Some perished there, while others
Were rescued by a rope;
Some leaped from the sixth story,
While many hearts were chilled,
To hear the groans of the dying
Around the Granite Mill.

I've seen that lonely Mother,
Beheld the falling tear,
I've heard her speak of Ellen,
Her only daughter dear,
Who in the flames did perish
Far up six stories high;
Great God console that Mother
And fit her for the sky.

From one of the attic windows
A lad sprang to the ground,
The lookers on expecting
In death he would be found;
He sprang up and ran home quickly,
And did these words repeat:
"Oh, Mother, the Mill's all on fire!"
And sank a corpse at her feet.

I've talked with that dear Mother,
Her heart with sorrow crowned,
That saw her darling daughter
Fall full six stories down.
None can describe the anguish
Which did her bosom swell,
None but the God of mercy,
Who doeth all things well.

Mourning as did poor Rachel,
She sighs at her hard lot,
Refusing to be comforted,
Because her girls are not.
Still weeping, sad and lonely,
Her heart with sorrow filled,
Bereft of her three daughters,
They died by the burning Mill.

The Agent of this Company
A noble deed hath done,
In helping the poor operatives
That worked in number one.
For which he'll be rewarded,
For all his good deeds bless'd,
With others who have aided
The wounded and distress'd.

A word now to the Engineers:
See that everything is right,—
Look well at your fire department,
At morning, noon and night.
See that everything's in order,
And keep in memory still,
The loss of life and suffering
By the burning Granite Mill.

"The Granite Mill Fire, at Fall River, Mass." Composed by A.W. Harmon, it is to be sung to the tune of "Young Edmond." Published in the *Fall River Daily News* in 1874, soon after the Granite Mill fire of that year. *Fall River Historical Society.*

A *Daily Evening News* editorial published the evening of the fire urged that safety should be made a priority in all mills, nationwide. The paper advocated that there be fire alarm bells on each floor of mill buildings, better inspection of safety equipment and mandatory worker fire drills to familiarize all employees with escape routes. The editorial wasn't far off from the changes that were soon implemented. In short order, the Granite Mill fire had prompted mill architecture safety and fire department reforms that would affect the entire nation.

The coroner's jury called for significant changes and safeguards to prevent the reoccurrence of such disasters. And as a result of the fire, laws and regulations were implemented to require that mill buildings have more than one exit. The law mandated outside fire escapes of metallic construction. And automatic sprinkler systems were afterward invented and placed into service. Legislative acts further made it compulsory for local governments to conduct periodic inspections of mills and other workplaces and to make certain emergency and safety devices were on hand and operational.[55]

Property damage for the Granite Mill fire was reported to be $247,000, but there was no financial loss to investors as the mill was fully insured.

Poets and songwriters created works to honor the dead and tell the story of the tragedy, many appearing in New York and Boston newspapers.

Eventually, the burned-out mill was gutted and rebuilt. The sixth-floor half-story attic was eliminated entirely, and the fifth floor was covered with a flat roof. The old clock stairway tower was modified to the more conventional arched belfry style, and additional stairways were added in the building. Years later, in 1962, the wreckers' ball crashed into the old granite bulwarks to make way for a supermarket/store/retail complex. Perhaps the ghost of Mrs. Murphy is finally at rest, and she no longer feels the need to bring breakfast for her three daughters who ceased to exist on that fateful day in 1874.

3

THE STEIGER STORE FIRE OF 1916

Great tongues of flame leaping upwards [sic] *until they seemed to lick the very darkness of the night, smoke belching heavenwards blacker than the night, it was a scene almost beyond the description of a master.*
—Fall River Evening Herald, *February 16, 1916*

In 1916, Fall River was in its heyday. The city's South and North Main Street business districts were thriving and bustling, crowded with department stores, theaters, small mom-and-pop shops, confectioners, grocers, wholesalers, professional offices (doctors, dentists, lawyers), undertakers, restaurants, churches, furniture stores, hotels, barbershops and hardware stores. According to the February 16 *Fall River Evening Herald*, Fall River was then the largest city in southern Massachusetts with a population exceeding 128,000.[56] The city boasted 111 cotton mills containing 4 million spindles and employing 37,000 operatives. The mills were weaving over two thousand miles of cotton cloth a day, including fine and course goods and ginghams, besides spinning yarn and thread. Investments in the mills exceeded $50 million and $805,000 was distributed weekly in wages.

The paper also touted other reasons why Fall River was worth visiting and a city in which to build a business:

- Large hat, piano, paper box factories and many narrow fabric industries
- Eastern terminal of Fall River Line, world renowned for equipment, spaciousness and beauty of decoration and construction

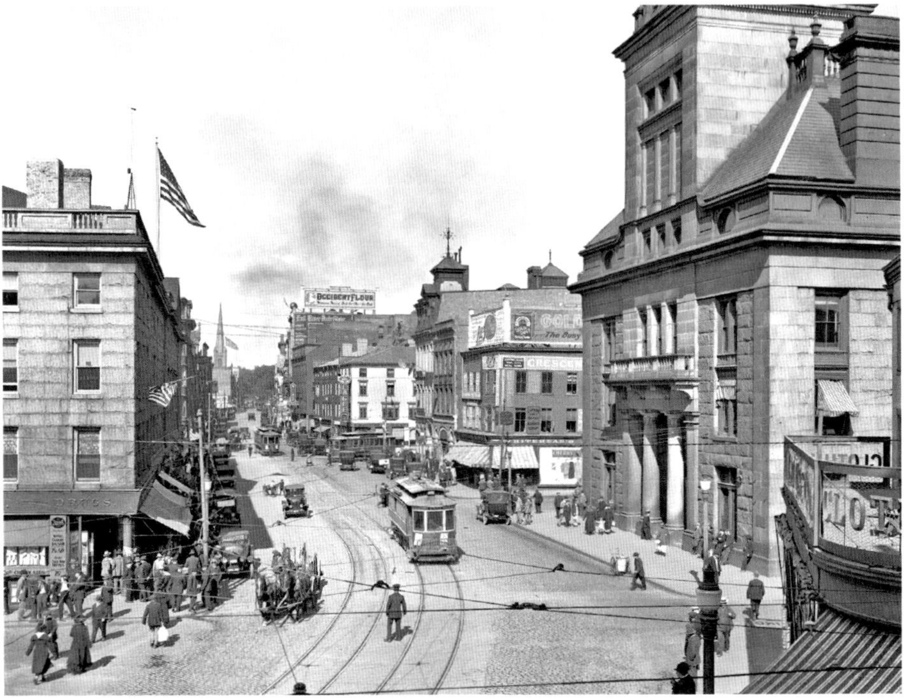

Main Street, Fall River, circa 1915. *Detroit Publishing Company Photograph Collection, Library of Congress Prints and Photographs Division.*

- On Old Colony Division of New York, New Haven and Hartford Railroad
- Fifty miles from Boston
- Unsurpassed harbor
- Eighteen miles inland from the Atlantic Ocean
- Natural market for prosperous agricultural district
- Practically inexhaustible supply of pure water for domestic and mechanical purposes, delivered by municipal water-pumping plant, with superior pressure in all districts
- Up-to-date fire department, equipped with motor and horse-drawn apparatus, manned by permanent firemen
- Eighteen miles of paved streets, 150 miles of accepted streets, 920 electric arc lights, 96 "White Way" lights, 709 other streetlights
- Adequate school system, with English Classical and Manual Training High School; seat of Bradford Durfee Textile School
- First free textbooks in America

- Churches of all denominations; YMCA; fine public library; clubs for men, boys, women and working girls
- Beautiful surroundings, between two bodies of water, exceptionally healthy community with climate unsurpassed
- Excellent location for manufacturing enterprises; with abundance of help, old and young, an attractive home site
- Fine stores owned by energetic merchants
- Abundant power, both steam and electrical
- Available space in buildings for leasing for small manufacturing enterprises

There was great discussion in the months and years before 1916 that South Main Street needed to be widened and buildings set back to accommodate streetcars, motorized vehicles and sidewalks. The road was just too narrow in its present state. The inability to come up with the funds and a solid plan to accommodate this necessary work was a continuing public debate. Editorials appeared in all three of Fall River's English-language newspapers advocating the expansion.[57] The mayor and aldermen met many times with local merchants who would be affected, but there never seemed to be enough will or money to make the improvements.

All of that changed on the night of February 15 and the morning of February 16, 1916, when the entire block framed by South Main, Spring, Pearl and Columbia Streets was completely destroyed, except for two buildings, in one of the worst fires in Fall River's history.

IT HAD ALREADY BEEN a terrible week for the city. Just three days earlier, on Saturday, February 12, fire had broken out in the engine room of Mill No. 1 of the American Linen Company. Incorporated in 1852 and located at the foot of Ferry Street, the cotton textile manufacturer was large by Fall River standards and had, according to its advertisement in the 1916 *City Directory*, 94,528 spindles and 2,610 looms and employed 390 men and 430 women.

The fire originated when a candle, or "open lamp," that an electrician was using while doing repairs in the engine room came in contact with heavy dust. It was reported that flames from the resulting explosion swept quickly through the mill's main belt box to the main pulley and then traveled into the tower and, through that, to the top floor of the mill.

Firefighters were on the scene quickly and directed their resources to extinguish the source of the fire in the belt box. While they were at work there, the intense heat from the blaze triggered the sprinkler system on the fourth floor and soon a downpour of water drenched the machinery and linen stock then in process, causing a great deal of damage.

The main pulley that operated the plant was over fifteen feet in diameter and was constructed of wood about eight inches in thickness. "Not only had the flames eaten the belts from it, but they had penetrated through the solid sections in some places and elsewhere materially weakened it."[58]

The loss of machinery and stock was estimated between $20,000 and $25,000, and at least half of the 820 operatives were thrown out of work as a result of the fire. Repairs were expected to take two to three weeks. In the meantime, almost four hundred people would have to make do, in this era before unemployment compensation, until their jobs returned.

Tuesday, February 15, had been a very cold day, with the temperature below zero by nightfall. The weather report called for continued frigid weather for most of New England, New York and eastern Canada. On this night, Northfield, Vermont, would record a temperature of negative eighteen degrees. It was so cold that the snow from the previous week's storm still blanketed the city and covered the roofs—not a fit night for man nor beast.

Near midnight, Ellis Goldberg of 189 Ridge Street, a forty-three-year-old real estate agent, was walking home with friends from the Young Men's Hebrew Association.[59] The group had just crossed Spring Street, reported Goldberg, "and just why I chanced to turn my head and look down that thoroughfare I do not know. Almost at the instant that I saw smoke which appeared to be coming from the side door of the basement, a little boy whom I do not know exclaimed, 'There's a fire, get the fireman.'"

Goldberg says he looked about for an open store that might have a telephone, but when he saw how thick the smoke was, he instead broke the glass on Box 412, located on the corner of South Main and Spring Streets, at 11:45 p.m., and sent in the alarm. "We all then ran down towards [*sic*] the side door of the building. After ringing the alarm I thought maybe I was a little too hasty in doing it, without making a proper investigation."

When he looked in through the basement window of the Steiger Store on the corner of South Main and Spring Streets, he was glad he had not delayed or taken any more time to call in the fire. "As near as I could tell you there was [*sic*] flames shooting up in or close to the southwest corner of the basement. Then our party went back to South Main street to watch the apparatus and wait around to see what the fire might amount to."

In four or five minutes, Goldberg said, the night watchman of the building appeared. "When he came out, he asked what was going on. I told him there was a fire in the cellar. He asked me where it was and I told him it was in the building he had just left. Then he ran back into the building."

Subsequent to the watchman going back into the building but unsure how long after, Goldberg said there was a sizeable explosion.

> *I don't believe I will ever forget it. It seemed that almost instantly after the explosion the flames shot from the basement clear up into the second floor. When they came out of the front windows on the main floor, and almost before you realized what that meant the flames were shooting out from the second floor. After that I heard what appeared to be a couple of other explosions. Now that I have seen what damage has resulted I was glad that I did not make a thorough investigation before I sounded that alarm.*

Goldberg's statement led state and local investigators to initially believe that a boiler explosion in the basement of the Steiger Store caused the fire. And as an innocent bystander, with no vested interest in the situation, his story rings true. Yet Goldberg's words also contradicted the night watchman's sequence of events about those early moments. We cannot know whether these discrepancies are the result of faulty reporting (most likely) or someone trying to cover up their negligence. It is interesting, nonetheless, to note that the complicated process used in determining the origin of a fire, especially in the early years of the twentieth century, most often involved eyewitness accounts of the premises before the fire and details about the fire itself, not expert examinations conducted by forensic fire scientists or fire investigators. One might even go so far as to say that in 1916 the use of science in the investigation of fire and arson was completely unheard of.[60]

Watchman Michael O'Rourke, fifty-seven, 110 Whipple Street, explained to reporters, as well as police and fire examiners, that his nightly routine at the Steiger Store always involved making a complete round of the store on the hour, every hour.[61] Besides what it took to complete the few tasks asked of him by the store manager, the time between O'Rourke's rounds was his own. O'Rourke said that he had been in the basement about 11:00 p.m. and noted that everything was as it should be. He registered his presence on the watchman's box located there. He then made his rounds, winding the watch clock on every floor. There are two different accounts about what happened next.

When O'Rourke reached the top floor, around 11:30 p.m., the *Fall River Evening Herald*, in a story published the next day, reported that he heard "a rumbling" and went to the basement to investigate. *The Standard*, a weekly insurance newspaper out of Boston published on February 19, has O'Rourke noticing "the strong smell of smoke" and, as he started for the lower floors, hearing "the crackling of flames."

In *The Standard*'s reporting, O'Rourke "immediately ran from the building to the fire alarm box at the corner of South Main and Spring streets to find that an alarm was being turned in by a citizen. About this time, the north windows of the building blew out with a tremendous crash."

In the *Fall River Evening Herald* story, when O'Rourke went to the basement to investigate the rumbling, "he was met with a blast of smoke and flame all but over-powering, and it was with difficulty he gained the street. He ran from the building to sound an alarm from the fire box at the corner of South Main and Spring streets only to find Ellis Goldberg, who had noticed the flames from the street, pulling the alarm. Watchman O'Rourke returned to the store with the intention of telephoning fire headquarters but was unable to reach the telephone."

Disputing claims of a boiler explosion as the cause of the inferno, O'Rourke said that when he first saw the fire it was "a hundred feet away from the boilers which are located in the southeast corner of the basement. I hold a fireman's license, and my first duty when I go to work at 6 in the evening is to look at the furnaces. When I looked at them Tuesday night, the fires were properly banked, and there was no rubbish or any material around the furnace which could readily take fire."

Asked if there had been any deliveries of goods recently where there might have been a large amount of excelsior or straw about, O'Rourke said that he could not say definitely for the whole building because he was not on duty during the day, but he was sure there had not been any excelsior or straw in the basement.

Having no theory for the start of the fire, O'Rourke called it "one of the most amazing things he had ever heard of, and marveled at the rapidity with which the flames spread. He considered himself most lucky to have escaped with his life."

The Steiger Store building was a three-and-a-half-story brick structure that also contained a full basement and sub-basement. The area of each floor was approximately ten thousand square feet. The store had just opened the year before and was still in the process of having its interior decorated with furniture, mirrors and displays. Steiger's had been doing such good business that there was a large amount of general department store stock on the premises. While there were no working sprinklers in the building, a

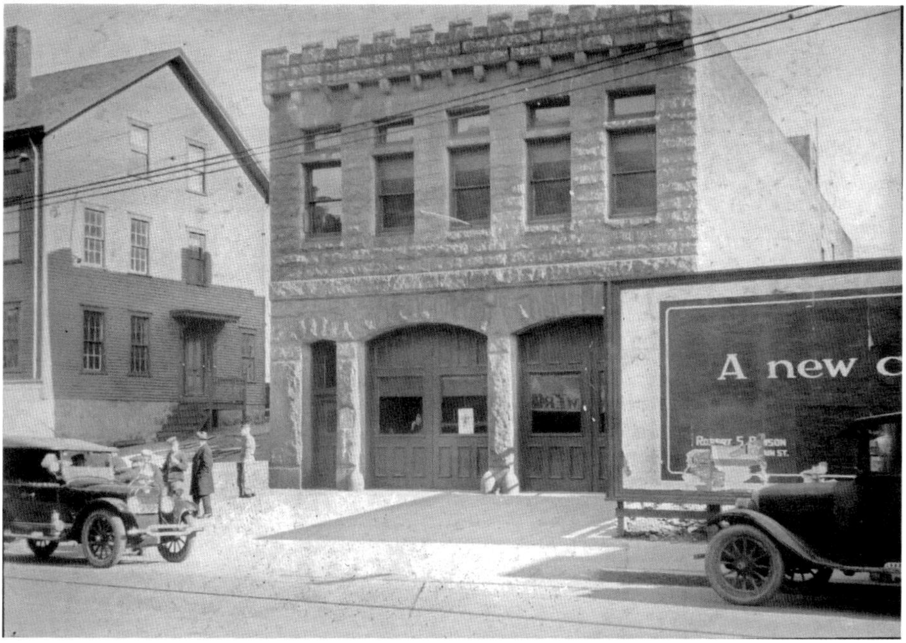

Cascade House, South Main, North of Morgan Street. The station is now part of the Goldstein & Horvitz Building, which housed the locally popular Modern Furniture for many years. *Fall River Historical Society, P.D. Borden Collection.*

complete system was being installed at the time of the fire but, unfortunately, had not yet been connected with the city's water main.[62]

Hose Company No. 8 from the Cascade House, located about a block and a half away at 384 South Main Street, between Rodman and Morgan Streets, was first on the scene. These firemen reported that the Steiger building was a furious mass of flames upon their arrival and that they hardly had time to hook up a line of hose before the windows of the Steiger Store on the Spring and Main Street sides had melted from the intense heat.[63]

Chief engineer William Davol arrived on the scene and assessed that the blaze was growing quickly enough to warrant a general alarm at 11:54 p.m. to bring out the full firefighting force of the city. But there was a problem. For some unknown reason, the alarm sounded only the four strikes, indicating the general alarm, but failed to follow with the number of the box, as is required.

Fearing that his general alarm might not be understood in the firehouses and cause a delay in the firefighters' arrival, Chief Davol drove to headquarters in city hall, two and a half blocks to the north, and directed Captain Jeremiah Sullivan, who would later serve as fire chief from 1922

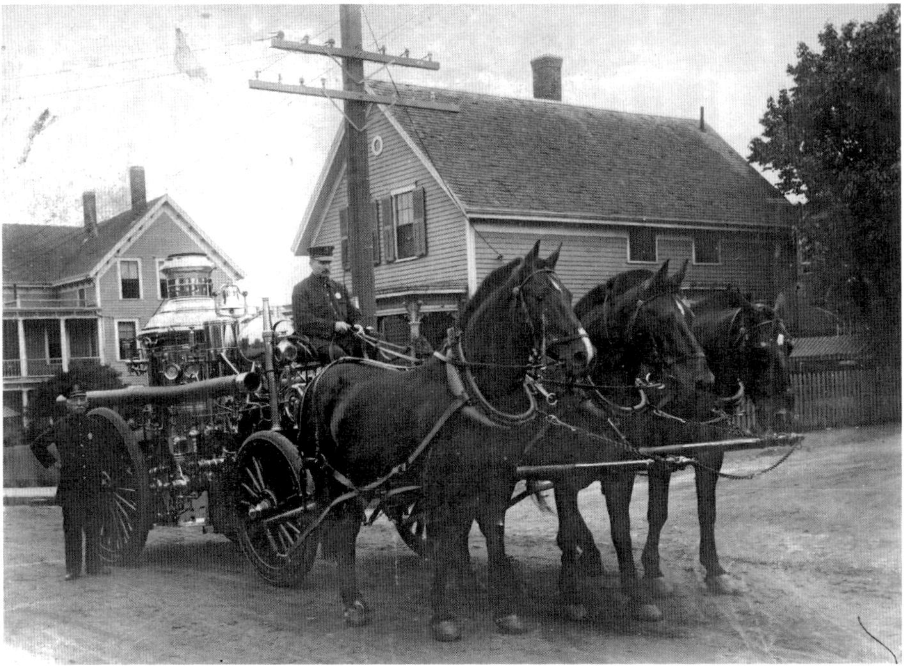

"Steamer Massasoit Before Steam Was Superseded by Gas." *Fall River Historical Society, P.D. Borden Collection.*

to 1937, during the Great Fire of 1928, to telephone all firehouses and make sure they were sending every one of their apparatus. Chief Davol then used the switchboard to call for assistance from the fire departments of New Bedford, Newport and Taunton. These calls were received only a few minutes after midnight. This quick thinking on the part of Chief Davol meant that assistance from these other communities would be prompt. In fact, not only did each city send two pieces of automobile apparatus, but the fire chiefs of these departments also came to assist in controlling the fire.[64]

Patrolman Harrington, who was a block away at the corner of South Main and Columbia Streets when the alarm sounded, said he had passed the Steiger Store only ten minutes before and saw nothing. Just as he arrived at the fire, the front windows of the store exploded with a tremendous crash and the flames jumped as far as across the street. When Steiger's north windows blew out, the flames jumped across Spring Street to the brick walls of the four-story Campbell Building, which contained the Kresge Five and Ten Cent Store, two jewelers, two dentists, an optician and the offices of the Knights of Columbus. The awning on the building was ignited, and the

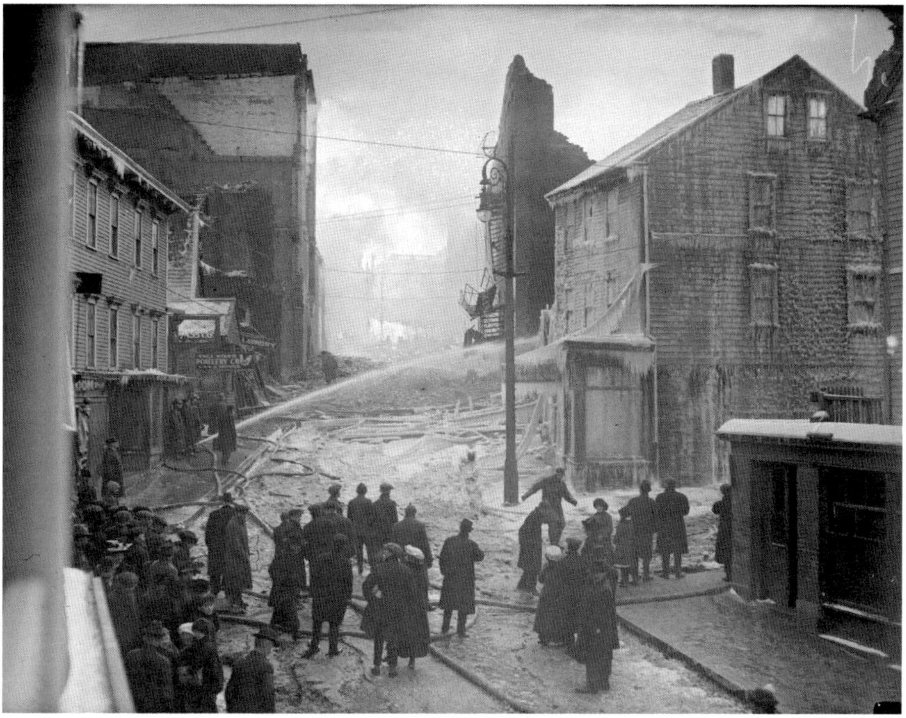

"South Main looking South, near Pearl Street," 1916. *Boston Public Library, Leslie Jones Collection.*

fire entered the structure through the windows. Simultaneously, the blaze had stretched to the buildings owned by Emma and Lizzie Borden on the southeast corner of Spring and South Main Streets—all four corners were now enveloped.[65]

The Campbell Building had been equipped with fire shutters, which were usually closed at the end of each business day. For some reason, on this night, they had been left open. "If these shutters had been closed there is little question that the fire would have been stopped at Spring street and the eight buildings to the North saved or at least a chance to work to advantage."[66]

With the fire spreading so rapidly, Harrington then made it his duty to alert anyone living or staying in the area about the threat of fire, first arousing the sleeping residents of the upper floor of the Eagan building, located directly across the street from the Steiger Store. It took several moments before the family realized the extent of the danger, even as their abode began to burn, and they had no time to leave with anything but the clothes on their backs.

Harrington next turned his attention to The Lenox, a forty-five-room, four-story brick hotel, also located across the street from the Steiger Store, on the east side of South Main Street. When he arrived, Mrs. Thomas Phelan, proprietress of the hotel, was already going room to room, calmly letting the guests know that the store across the street was burning and not to be alarmed if they smelled smoke. According to M.E. Burbank of Boston, a guest at The Lenox that night:

> *All of us got up, of course, and inside of 15 minutes the entire front of the hotel was ablaze, and in less than an hour the building was flat on the ground. We got out by means of fire escapes at the rear; every one was orderly and the most attention was given to three babies in arms, who were known to be in the hotel with their parents. Everyone seemed first to think to get the children out safely. Everything was sidetracked for the babies.*[67]

Several other people also rushed into The Lenox to warn the lodgers, including Ormon Bassett, Benjamin Read, Thomas Powers and William Murphy, who carried the children of the proprietor, Thomas Phelan, down the fire escape. There had been no time to dress them, so "they were hurriedly wrapped in blankets and hustled out of the building."[68]

John Rogers, a twenty-seven-year-old bookkeeper at the Pardee & Young Coal Company, went to The Lenox to make sure his friend, Clinton Smith, a guest at the hotel, was safe. By the time he was pounding on Smith's door, firebrands had crashed through the window of the room and set the curtains ablaze. As Smith struggled to put on his clothes, they caught fire. Rogers helped his friend to extinguish the flames, but not before they had been scorched in several places. It was a close call.

A blind man, who was a long-term roomer, required assistance in escaping that night. As he staggered down a corridor in The Lenox, "some one noticed the outstretched arm with faltering fingers feeling along the wainscoting. The next instant willing hands had grasped either arm of the blind man and he was being led along the corridor." They had a bit of difficulty getting him down the ladder of the fire escape, but he did make it out safely and was taken to a "place of shelter."

According to the *Fall River Evening Herald*, almost all of The Lenox Hotel's fleeing guests worked in the theater or were their family members, including Booth & Leander, Nina the banjoist, Hunt & Read and other widely known vaudeville people. Members of the French Model Theatrical Company, a burlesque group in town playing at the Academy Theatre, lost all of their belongings, including clothing, jewelry and personal effects, as they escaped into the night. Their costumes and props were saved only because they were securely stored in the theater, which was a few blocks from the fire district.

Because the fire spread so quickly to The Lenox and the hotel burned down with such rapidity, rumors began circulating that many lives had been lost there. The crowd that was gathering to watch the events of that night believed that four children of vaudevillians had died in the flames. To everyone's relief, it was not so. Everyone staying and working at The Lenox Hotel made it out safely.

Manager Christopher Clark of The Mohican, located at 6 North Main Street, opened his doors to those fleeing The Lenox. His hotel was already full, but his guests were only too happy to lodge anyone who needed a place to stay. The Mohican ended up caring for twelve of The Lenox roomers with the remainder finding other accommodations. It was reported that a wife of one of the performers appeared at The Mohican with her willow-plumed hat slowly smoldering away. She had not even been aware that this was happening.[69]

Perhaps an understanding of the astonishing velocity of the flames as they spread from building to building can be gleaned from experiences of occupants of the Campbell Building.

> *Several young men were preparing to lock up the rooms of Fall River Council No. 86, Knights of Columbus, when first aware of the fire in the Steiger building. Viewing the fire for a few minutes from the window, they decided that there might be some trouble so the pool table and some other fixtures were moved back from the windows. The danger, they thought then, was probably more to be expected from water and smoke than from fire. Hardly had they got out of the building when the quarters of the Knights of Columbus was in flames and instead of property being scorched or damaged, as they feared might happen, everything was destroyed.*[70]

Within twenty-five minutes from the first fire alarm, the Steiger Store collapsed into a mass of ruins, amazing the firemen with "the very brief time in which it took the flames to wreck this modern department store." They had been shooting streams of water into the structure from every angle but this effort had not the slightest effect in combating the raging inferno. "The collapse of floors and other interior sections gave the firemen some warning that the entire collapse of the building might be expected momentarily so that ladders were cleared away and apparatus was moved." There were no

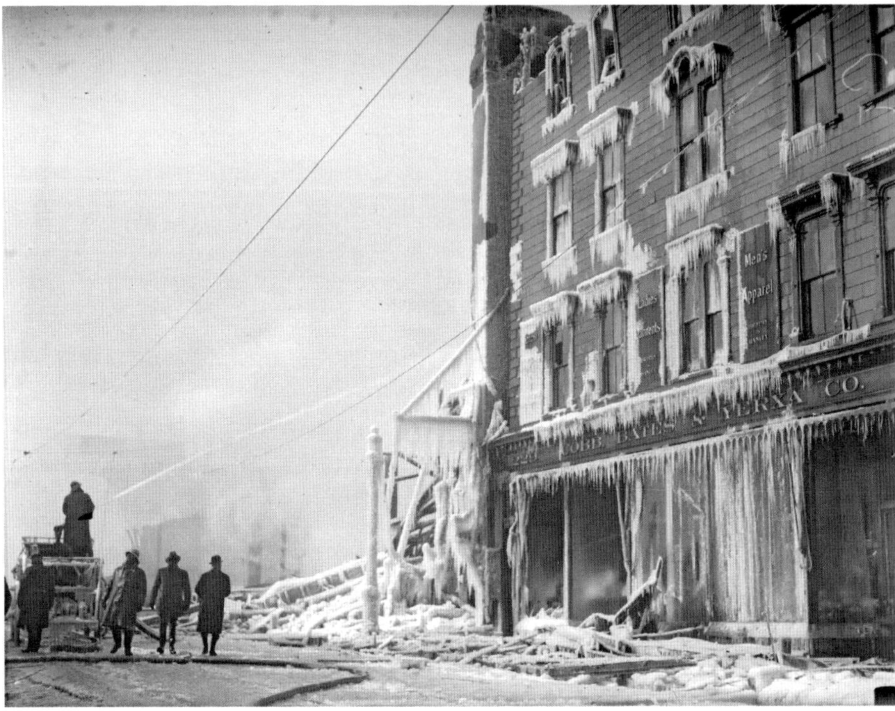

The ruins of South Main Street the morning after the Steiger Store fire of 1916. At the beginning of the twentieth century, the Cobb, Bates & Yerxa store, a Boston-based regional grocery chain, was believed to be the largest grocery retailer in New England. *Boston Public Library, Leslie Jones Collection.*

injuries among the firemen because they had pulled back when they sensed the building's imminent demise.

"Flying timbers and brands from the collapse" of the Steiger Store helped to expand the fire area and very shortly "the Burgess and Sullivan buildings, adjoining the Steiger store, were enveloped in flames."

Firefighters poured tons of water into these buildings with little or no effect. After the wooden structure of the Eagan property caved in, The Lenox collapsed. "The scattering of the debris from all these buildings added still further to the spread of the fire and on both sides of the street buildings to the south were ablaze and rapidly getting beyond the control of the firemen."

Because of the increasingly uncertain nature of this conflagration, Chief of Police William Medley called on every man in the force to report for duty. Some were ordered to go house to house in the Pearl, Second and Third Street neighborhoods and advise families to pack their belongings and prepare to leave their homes if the situation warranted.

> *Men, women and children fled with household goods toward the water front. Push carts, wheelbarrows and everything that could be pressed into service to carry things were utilized, the procession of refugees being scantily clad, some barefooted. Several sick people were carried from their homes, in some instances stretchers being utilized, and in others the beds were taken up bodily. From one house the body of a man who had died the day before was carried to an undertaking establishment.*[71]

Meanwhile, to the south of the Steiger Store, after the raging inferno had destroyed the small wooden building owned by tobacconist Frank Burgess, firemen were able to get into the next structure, the Sullivan Building, which housed Sullivan and Sons undertakers, with several lines of hose. They soon had to beat a hasty retreat. This wooden structure also furnished ready fuel to the fire. "It collapsed like the others, leaving nothing but a small part of a partition standing."[72]

Farther south, the brick Flint Building, occupied by Cherry & Webb, offered no more substantial resistance than any of the wooden structures on the street. When Chief Davol was certain that this building would also be lost, he decided that they would make a decisive stand at the Edwards Building, two doors down, on the corner of South Main and Columbia Streets.

When the fire had first commenced, Chief Davol stationed apparatus and firefighters to completely surround the block in which the fire was raging. At the Columbia Street end to the south, Deputy Bowers and his men worked to hold the fire in check at the Edwards Building, and very slowly, it became more and more certain that the plan had worked. On the east side of South Main Street, the fire had spread from The Lenox Hotel to the next succession of wooden buildings, quickly consuming them. When the fire reached the Manchester house, located near the corner of South Main and Rodman Streets and "occupied by tenants up stairs and a shoe-shining parlor on the first floor, the firemen, encouraged by the hold obtained across the street, put forth even greater effort and here, too, the flames were held in check."

To the north of the Steiger Store, a heavy firewall between the Campbell Building, on the corner of South Main and Spring Streets, and its next-door neighbor Cobb, Bates & Yerxa (CB&Y) served to noticeably check the spread of the fire. Chief Davol knew that if the fire extended to the McGuire Store (men's furnishing goods), the next building north of the CB&Y at 209 South Main Street, there would be very little chance of saving that entire stretch of structures as far down as Anawan Street and perhaps beyond. He directed an immense volume of water from multiple streams into the CB&Y from the

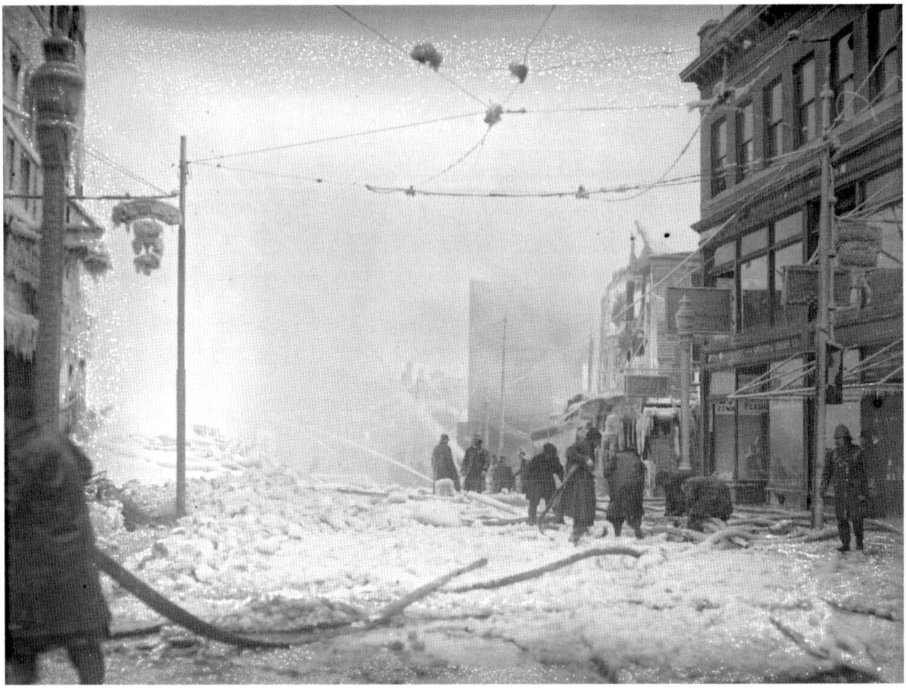

South Main Street on February 16, 1916. Firemen are attempting to dislodge the frozen hose lines used to fight the fire the night before. *Boston Public Library, Leslie Jones Collection.*

roof of the McGuire Store as well as from the building's front and rear. Although the roof was destroyed, the floors were caved in and the walls were gutted, this slowing of the progress of the inferno gave the firemen a chance to get a foothold that enabled them to soon control the spread in this direction.

The R.A. McWhirr Company, located two buildings north of the McGuire Store, at 165–93 South Main Street, had been equipped with an exterior sprinkler system that had been turned on early in the conflagration, preventing sparks from igniting the store. Only a small amount of water had found its way inside from the heavy flow from the sprinklers, and it amounted to only minor damage. In fact, the store was open for business the next day.

In spite of the fact that it was located just north of the McGuire Store and was under threat of being included in the conflagration, the Baptist Temple "turned itself into a sort of field hospital for the fire workers." The two large basement furnaces were lighted and anyone who was in need of warmth was welcomed into the church. The Sunday school room of the temple had a kitchen, and it was turned into a sort of "coffee tavern, and young ladies connected with the church, heedless of insistent orders to seek the comfort of home, carried the welcome hot drink in steaming cups right onto the line of attack." Exhausted firemen, coated in ice, "tramped their way through the immaculate corridors of the quiet church when fleeting opportunities offered, and almost in the glare of the flames from the furnace-like wooden building to the south sought momentary comfort and warmth."[73]

Other local businesses likewise opened their doors to the firemen, offering hot drinks, sandwiches and smokes, including the St. James Café at the corner of South Main and Columbia Streets; Edward W. Cantwell, druggist, on the opposite side of South Main Street; Kirby and Baker pharmacies; and the office of Dr. J.M. Leonard. Harrington Brothers' clothing shop offered the firemen and the fire's refugees gloves, socks, mufflers and other wraps. Clark Chase Jr. loaded cans and mugs of hot coffee into his car, and the beverages "were carried inside the firelines to the ice-coated men."[74]

It was feared that if the Morris Paint Store, at 285 Spring Street, caught fire, the flammable materials stored there would destroy St. Mary's Cathedral, its neighbor to the east, so great effort was made to protect the granite church and its holy contents. During the height of the blaze, sparks had lighted on the slate roof of the church and encircled the steeple. The rector, Reverend Francis Bradley, and his assistants removed the sacred vessels as a precaution.

It did not take long for the paint shop to become a mass of flames. In only few minutes, nothing but ruins remained. Luckily, only a few slight explosions occurred inside the building, and none of them was large enough to spread brands beyond the property. With the store's destruction, "the greatest danger that menaced the church had passed."[75]

On the north end of the fire zone, storeowners, aided by scores of willing assistants, began the formidable job of removing the entire stock of their stores to save their goods, transferring items by the armload to backyards, residences, other stores farther away and even automobiles. The massive quantity of water that had been pouring over the sidewalks and streets had frozen almost instantly in the subzero temperatures, making this arduous task even more difficult.

Finally, at 3:00 a.m., Chief Davol announced that the conflagration was under control. There was still considerable fire at CB&Y to the north and Cherry & Webb to the south, but the buildings in between were mostly leveled. The danger of the blaze spreading farther had been greatly reduced.

Apparatus was left on the scene "in readiness for immediate use if the wreckage and debris, which burned slowly most of last night, should break

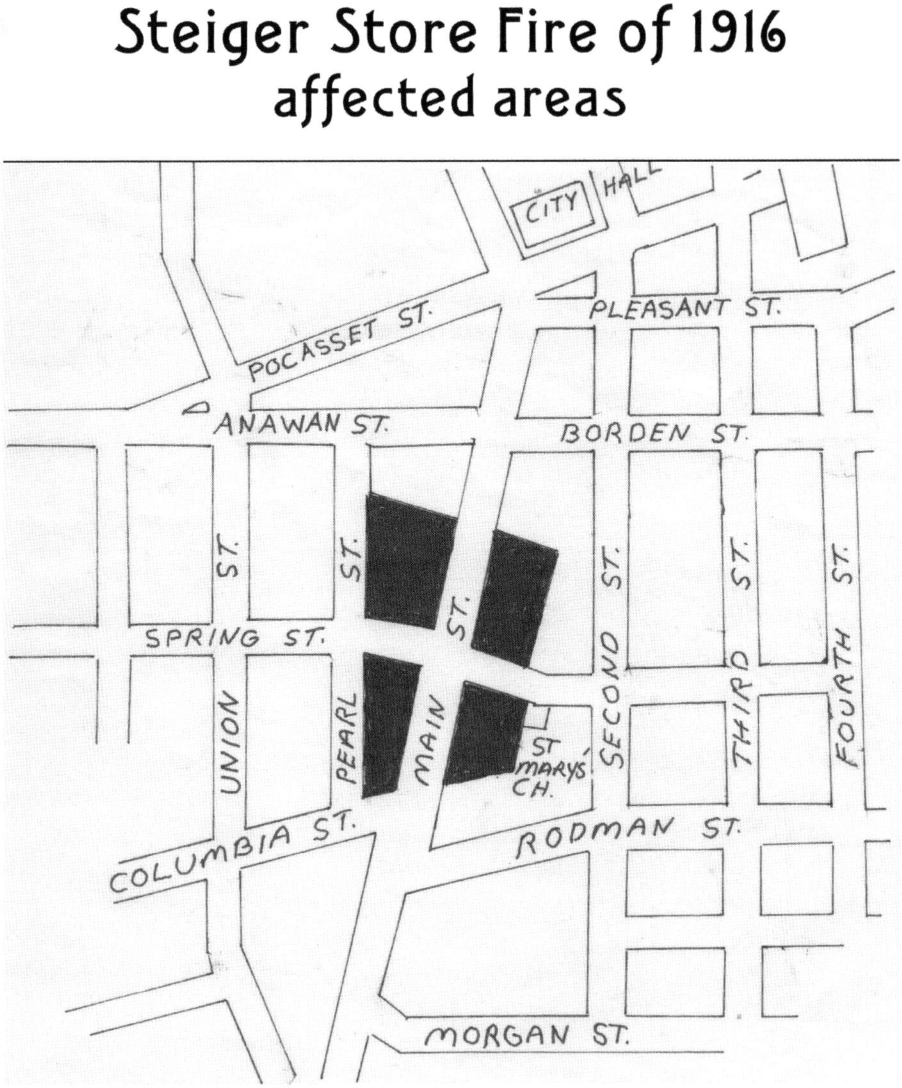

Diagram of the burned district in the Steiger Store Fire of 1916. *Adapted by the author from* Fall River Fire Pictures, *City News Company, 1916.*

out again. The lines were withdrawn primarily to permit the Street department forces to attack the mammoth proposition of removing the tons of bricks, plaster, twisted from and charred timber which effectually blocked all traffic."

Shortly after 10:30 a.m. on Thursday, February 17, Hose 3, whose powerful nozzle had been in continuous play since 12:17 a.m. on Wednesday morning, was shut off, and the last of the streams, which had been running all night directly from the hydrants and under city pressure, were withdrawn, completing thirty-four and a half hours of continuous service.[76]

All told, about fifty buildings, including some of the largest business blocks in the city, had been destroyed.[77] An examination of assessors' maps showed that the fire had traveled roughly 325 feet along both sides of South Main Street and slightly more than 200 feet on Spring Street. The fire affected only two tenement houses: a small roof fire occurred at the house of J. Mossoff at 133 Washington Street, and the house located at the northeast corner of Spring and Pearl Streets, directly adjoining the Steiger building to the rear, was badly gutted.

In a fire this large, it is amazing that there was no loss of life, and only minor injuries were reported by the firefighters. It was reported that during the fire several lines of hose were destroyed in the vicinity of Spring Street, but the department's equipment loss was quite small considering the work they had performed.

City workers soon began the laborious task of cleaning up the burned district, removing massive amounts of debris from South Main Street. In one day, they had accomplished a great deal and "had begun upon the task of restoring the sidewalks for use. One of the most difficult tasks was the removal of iron girders of the Steiger building which had melted and bent over into the street, their bases still firmly imbedded in the original foundations."[78]

The day after the fire, the Savoy Theatre advertised "Fall River Fire in Motion Pictures" in addition to "Keystone 3 Reels." Camera operators had been busy recording the conflagration and had driven the footage to Boston in the morning for processing, returning with it for the show that night.

Talk began almost immediately after the threat had passed to take advantage of the destruction on this section of South Main Street to incorporate the widening of the street into the plans for reconstruction. The *Fall River Daily Globe* urged the city's elected officials to take this action in an editorial published on February 16, while the smoke was still rising from the frozen ruins of the downtown business district and continued the drumbeat until it was a certainty.

A week later, twenty thousand people had visited the site of the great fire. Police captain Baker had ten policemen on duty and stationed in the area to make sure that "the over-curious did not risk their lives by driving into the ruins at any point." He reported that people thronged to the scene all day long "until it seemed to constitute the principal diversion of the holiday." Baker estimated that "by nightfall there was no one in Bristol county who

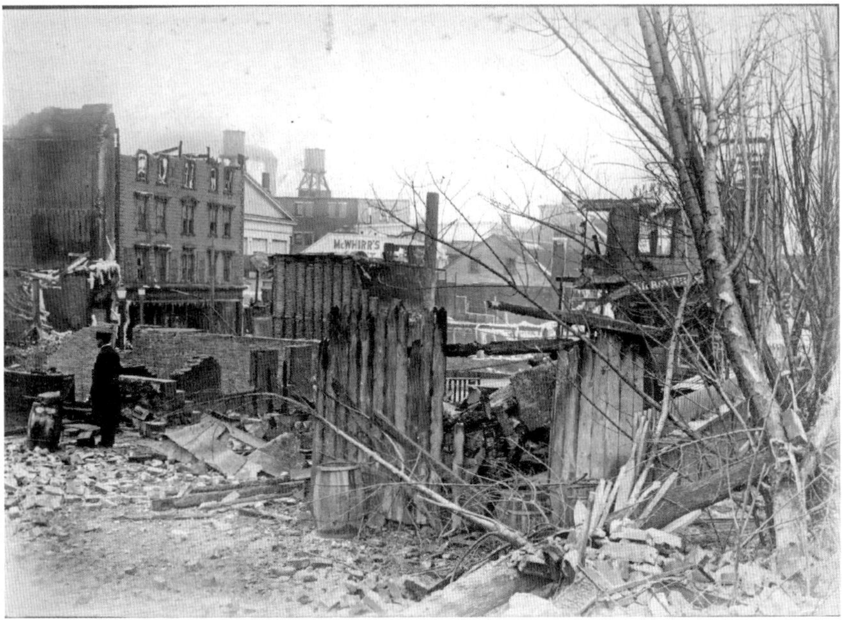

"Fire of Feb. 15–16, 1916, from Spring Street, North." *Fall River Historical Society, P.D. Borden Collection.*

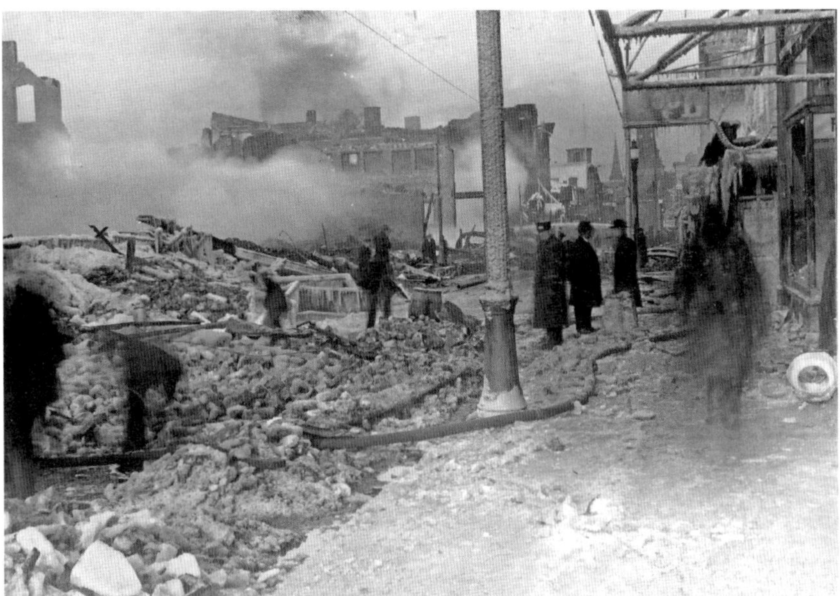

"South Main Street, South and North of Spring Street," 1916. *Fall River Historical Society, P.D. Borden Collection.*

had not visited the scene, and that the number of useless and frequently foolish questions that the policemen were obliged to answer was no less than 9,768,246." It was suggested by a bystander that "the city could have covered a large part of the expense of widening the street if the spectators had been charged 10 cents apiece for admission into the fire zone."[79]

Several positive factors were at play in this conflagration that worked to lessen the extent of Fall River's structural damage. When the fire had first been underway, there had been a fairly strong southeast wind blowing, but within an hour, it had let up, which aided the fire department greatly. Additionally, snow from a recent storm covered roofs in the city and served as a natural protection for the dwellings throughout the residential neighborhoods of Second, Third, Whipple, Cottage and Pearl Streets.

> *The snow covered roofs...practically saved the big central section of the city, as burning brands as large as a man's leg were carried for a half to three quarters of a mile to alight on shingled roofs, where they were smothered in the snow. There was also a thickly-built tenement district closely adjoining the fire swept section and nothing prevented the destruction of this property but the snow covered roofs.*[80]

Following the fire, affected companies made plans for the resumption of services. Cobb, Bates & Yerxa began work in new quarters in the Jennings Building, 103 Pleasant Street, but by 1926, they were no longer in business. The Steiger Company, the New York Cloak & Suit Company, S.S. Kresge's Five and Ten Cent Store and The Fashion wanted to rebuild their establishments in their same location.

Albert Steiger, founder and president of the Steiger Company, with flagship stores in Holyoke and Springfield, Massachusetts, arrived from Boston about 3:00 a.m. during the blaze and set up his office in the Hotel Mellen. Interviewed the next day, he stated, "In all the years I have been in business I have never had a fire originate within my walls. We have often suffered small losses from fires of outside origin but never within my store." With a reputation built on goodwill and customer service—"Your money's worth or your money back"—he personally vowed to remain in Fall River and rebuild.

> *You can tell the people that I am going to stay here and rebuild. If you had not treated me well this would have been a good chance for me to get out. You all have treated me well, and I am going to stay, and build a better store*

> *than I had before. Don't forget to emphasize the fact that the fire will have one big result that will be good. I mean the widening of South Main street. That has been greatly needed and the city authorities will certainly insist upon its being done.*[81]

And rebuild he did. The store in Fall River remained for another twenty years, closing its doors in 1927. Albert Steiger died in 1938, and his family continued the business until 1994, when the Steiger Company closed four and sold the remaining six of its stores to the May Company, the largest department store chain the United States, for an undisclosed sum.

It is important to note that most of the burned area was included in the "congested value" district as defined by the National Board of Fire Underwriters in a special report on Fall River issued in the year before, in June 1915. The report pointed out that the block bounded by Main, Columbia, Pearl and Spring Streets formed an excessive fire area with a high-potential hazard; only slight attention had been given to fire protection, with few automatic sprinkler systems installed. Prophetically, it was this block that was completely destroyed.

Following an inquest in which eleven witnesses were called, Deputy Chief George C. Neal of Boston and members of the Massachusetts district police issued a determination that the fire "was of mysterious origin" and that "there was no reason to suspect any one of an ulterior act, and that the possibility of incendiarism [*sic*] was quite removed. Unless of new facts, the findings of the deputy chief ends State action in the case."[82]

At the inquest, it was discovered that "electricians were employed all day Tuesday in laying wires in the sub-basement, and also that much of the woodwork in that section had been newly re-varnished. The fire might have started," said Deputy Neal, "either in the basement or the sub-basement. The fire was not caused either by a boiler explosion or defect or by a gas explosion," he said.

The Steiger Store Fire, as it became known, will be remembered not only for a financial loss that amounted to over $1 million but also for what it wasn't—a human tragedy with fatalities. It was a conflagration that started without warning and was caused by some unknown event. For a city with experience in dealing with unsolved mysteries (Lizzie Borden), this conundrum was a familiar state of affairs.

4

THE GREAT FIRE OF 1928

This is the greatest calamity that has ever visited the city.
—Mayor Harry Monks, February 3, 1928

The fire raged for more than eight hours. Three hundred and forty-eight firefighters and forty-one apparatus from twenty-two cities and towns, as far away as Boston, worked in zero-degree temperatures to halt the conflagration. Damages were conservatively estimated to be $35 million. Thirty-six buildings were leveled, destroying 261 individual businesses in six square blocks in the heart of the city, including three hotels, seven banks, two theaters, one newspaper building and several large business blocks.[83] One thousand people were instantly unemployed. And while 250 people were hospitalized, 20 for overnight observation, and some firefighters suffered frostbite and broken bones, there was, miraculously, no loss of life.

Fall River's Great Fire of 1928 was an accident waiting to happen. Like the aging business districts of so many New England cities, the burned area of downtown Fall River was crowded with old buildings of inferior construction, built before the age of more modern fire-preventive practices like fire division walls and automatic sprinkler systems. Streets that were fifty to seventy-five feet in width were no defense against the reach of flames fanned by a strong wind. It could easily have been predicted that once a blaze got started here, it would sweep unchecked, with the greatest potential of becoming a full-fledged conflagration. In fact, the properly protected buildings equipped with wired glass windows, tin-clad shutters,

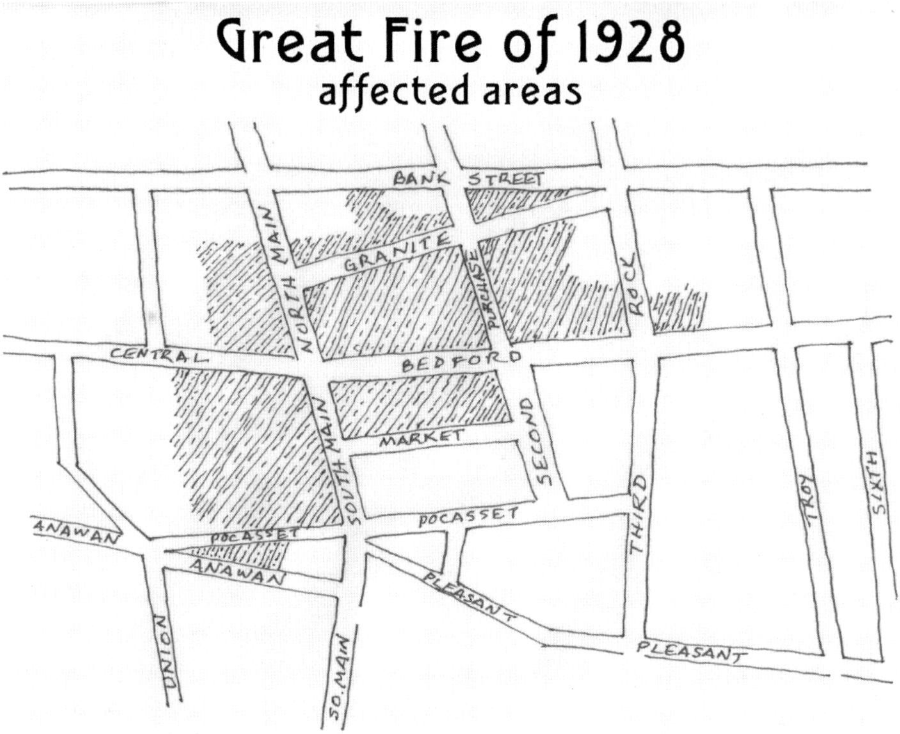

Diagram of the burned district in the Great Fire of 1928. Adjoining buildings on all sides of the fire area suffered smoke and water damage. *Adapted by the author from* Pictures of the Worst Fire in Fall River's History, *Reynolds Printing, 1928.*

open sprinklers, automatic sprinklers adjacent to windows and firewalls were instrumental in halting the spread of the inferno in several directions.

About 6:15 p.m. on the evening of Thursday, February 2, 1928, George Sullivan, circulation manager, and William Cooper, his assistant, were just leaving work at the Fall River Herald Building. A cold westerly wind was roaring up Pocasset Street as the men braced themselves against the icy zephyr. They happened to glance across the street, and both men noticed a flickering light in the windowless Pocasset Mill No. 2, just one of five voluminous structures between Pocasset and Central Streets that were currently in the process of being dismantled. They knew no one was supposed to be there working at this late hour. The light grew more radiant, signaling the escalation of flames. They quickly reported what they saw to Patrolman James Galvin, who was walking his beat in the area. Galvin sounded the alarm at 6:27 p.m. from Box 1253, located on the corner of Pocasset and South Main Streets.

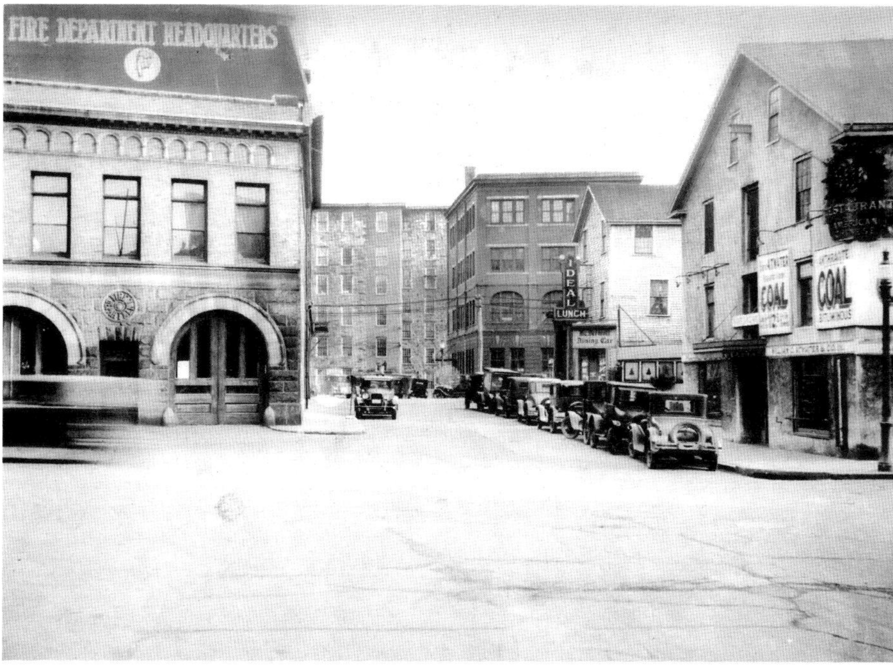

Pocasset Street, East of Second Street; Central Fire Station on the East Side of Second. *Fall River Historical Society, P.D. Borden Collection.*

The Central Fire Engine House was located behind City Hall only about eight hundred feet away, but even in the short amount of time it took the firefighters to respond, the smoke had become so thick that they could not locate the origin of the blaze. The fire grew more rapidly still, reaching a belt tower that connected with all the floors of the mill, granting the flames unrestricted access to the entire building, and beyond the control of one fire company. The second alarm was given as 6:34 p.m., and fire chief Jeremiah F. Sullivan went to the nearby newspaper office and called back to fire headquarters to signal a general alarm, bringing out all the available apparatus in the city and requesting calls be made for assistance from nearby cities and towns. By seven o'clock, three of the Pocasset Mill structures were burning ferociously. This was just the beginning of a very long night. And it would be another four days before the last of the fire apparatus would be removed.

The Pocasset Mill complex, a group of old cotton manufacturing buildings, was located behind the Granite Block, between Central and Pocasset Streets, on the west side of Main Street, across from City Hall. The mills were in the process of being demolished, and the upper floors of Mills Nos. 4 and 5 had already been taken down and holes cut through the floors for the purpose of dismantling machinery. In the months before, crews had removed the high-value salvage metal from all of the Pocasset Mill structures—plumbing fixtures, heating equipment and piping, including, regrettably, the mill's automatic sprinkler system, windows and fire doors. Without sprinkler protection, in fact, the vacant Pocasset Mills were a time bomb. Decades of machine oil had seeped into the wide board pine floors, creating a kind of grotesque tinderbox. If there had been careful day and night supervision of the structures, watchmen might have noticed the burning embers in an improvised salamander,[84] made from a fifty-gallon steel oil drum resting directly on the floor of the first story of the engine room of Mill No. 2, which workers had used during the day to keep warm. If so, this narrative would have had a happier ending. As circumstances dictated, however, there was nothing, mechanical or human, to stop the fire from growing out of control and totally destroying almost everything in its wake.

THE POCASSET MANUFACTURING COMPANY was the third textile concern incorporated in Fall River. Founded in 1821 by a group of New Bedford investors led by Samuel Rodman with $100,000 in capital, its original purpose was not to manufacture textiles but to provide rental space for industry. Known as the Old Bridge Mill, this three-story stone structure was located on the seventh falls from tidewater,[85] west of Main Street, and measured one hundred by forty feet, with a long ell on the south end extending over "the stream," what early residents euphemistically called the Quequechan River. The Hawes, Harris Machinery Company occupied two stories of the building, and Oliver Chace operated a gristmill, a mill for grinding grain, in the basement.[86]

A second mill building was constructed in 1825 west of the ell. John and Jesse Eddy occupied the southern half and manufactured satinets, a finely woven fabric with a finish resembling satin, and because of this, the mill became known as the Satinet Factory. Cotton textile printing began in Fall River when Andrew Robeson, related by marriage to Samuel Rodman, established his first calico printing business in part of the same building. Using the hand-and-block technique, his space contained one hundred tables for block printing.[87] The first textile printing machine built in the United States was used by Rodman and was constructed in the Pocasset Mill—a roller-printing machine made by Ezra Marble (1807–1890) of Somerset, Massachusetts, and a French immigrant who had knowledge of the machinery of Europe.[88]

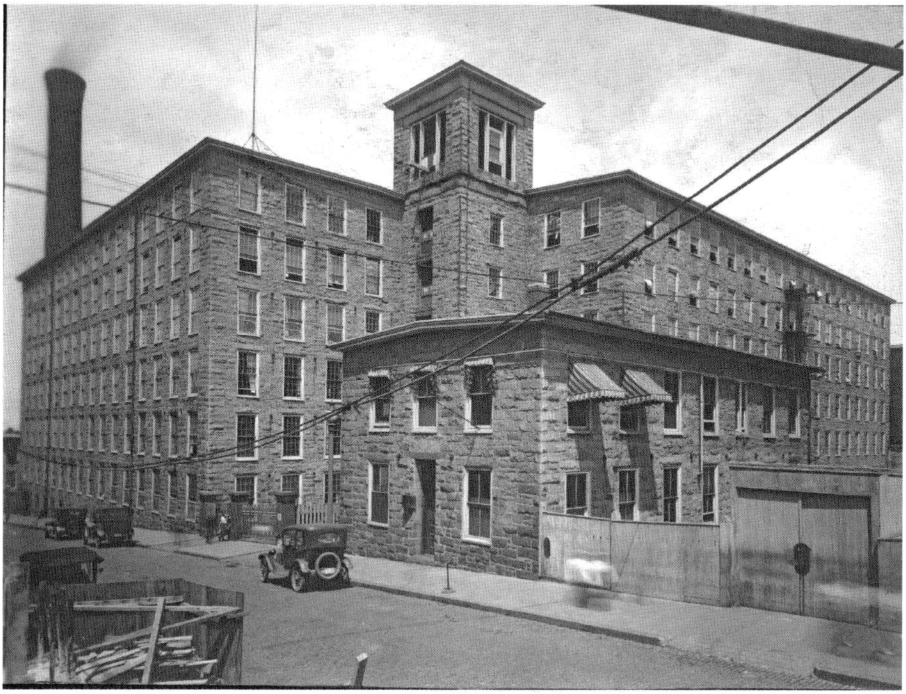

"Pocasset Mills in 1926." This was the site of the origin of the Great Fire of 1928. *Fall River Historical Society, P.D. Borden Collection.*

In 1826, the five-story, pitched-roof Quequechan Mill, called the "New Pocasset," was built west of the printing mill, across the sixth falls.[89] Built of stone and measuring 319 by 48 feet, it was leased for the manufacture of cotton yarn and cloth and contained 16,392 spindles and 392 looms.

In the following year, the Pocasset Manufacturing Company constructed still another stone building, across the fifth falls,[90] which was known as the Massasoit Mill and later called the Watuppa Mill. It was three times larger than any mill built previously so was considered too large for just one firm to occupy the whole of it. A partition wall was constructed from the foundation to the roof with two wheel pits so that several companies could occupy the space. Holder Borden, however, leased the entire mill in January 1831 and filled it with nine thousand spindles. It was successful from the start. He manufactured sheetings, shirtings and other fabrics. In this mill, Borden set up the cloth-printing business that would later become the American Print Works.

The Old Bridge Mill of the Pocasset Manufacturing Company was destroyed in the Great Fire of 1843. The first building constructed by the company

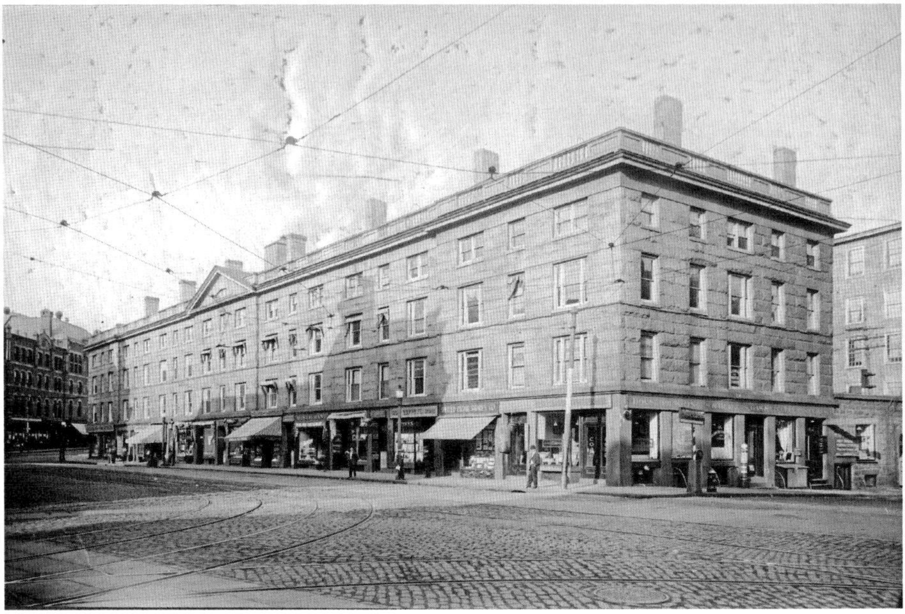

"Granite Block, Central to Pocasset Street." This entire block was burned to the ground in the Great Fire of 1928. *Fall River Historical Society, P.D. Borden Collection.*

following that conflagration was the four-story Granite Block, which was to be a center of commerce in Fall River for the next eighty-four years.

Four years later, in 1847, the five-story Pocasset Mill No. 1 was built on the site of the Satinet Factory. Mill No. 1 was constructed of granite and measured 208 feet by 75 feet, with a pitch roof containing thirty-two dormers and a prominent square tower on the south end. The dormers on the top were later flattened to a full sixth story. This mill manufactured shirtings and sheetings, and by 1877, the company employed 550 operatives and had built fifty-four tenements to house its workers.

Pocasset Mill No. 1 was the first of the so-called wide mills and ran 20,352 spindles and 422 looms, operated by three turbine wheels, later supplemented by a Corliss engine fed by a steam plant with eighteen condensers. With the loss of the Old Bridge Mill to fire still fresh in the mill owners' minds, Pocasset Mill No. 1 was equipped with its own fire apparatus: two force-pumps, stand pipes, hydrants, sprinklers and complete connections with the city's water works.

The Quequechan Mill was demolished in 1880 to make way for the construction of Pocasset Mill No. 2, on Pocasset Street; Pocasset Mill No. 3, on Central Street; and a new power plant. The Pocasset Manufacturing

Company now ran 56,112 spindles and 1,385 looms on sheeting, shirting, drilling, duck and print cloths; employed 730 hands; and had a capital of $1,161,000. By 1889, a complete electric light plant had been installed.

The Watuppa Mill was later refurbished as Mill No. 4 and ran 9,000 spindles and 224 looms. In 1905, the Fall River Manufacturing Company was acquired and became Mill No. 5. By 1917, the Pocasset Manufacturing Company operated 123,000 spindles and 2,874 looms, with a capitalization of $1.2 million, producing sateens (cotton fabric woven like satin with a glossy surface), twills (fabric woven to have a surface of diagonal parallel ridges) and plain cloth. Only nine years later, in 1926, the entire complex was shut down when the Pocasset Manufacturing Company ceased operations.

The first wave of mill closures in Fall River occurred in 1923. The end of World War I in 1918 brought a brief increase in manufacturing, but that postwar boom quickly slowed. The textile mills in New England faced serious competition from southern manufacturers with their lower labor and transportation costs. Southern textile companies had wisely reinvested profits into modernizing their manufacturing process, buying new machinery and other equipment, which additionally drove down their costs. Mill corporations in Fall River, ever manic about the bottom line, chose not to reinvest into new technologies or increase their capacity to compete with their southern counterparts and quickly fell behind.

After nearly one hundred years of mill expansion, it was a sign of things to come that only one new textile mill was built in Fall River after 1912. In 1922, the price for print cloth seriously declined, and by the next year, employment in the mills was down by one-third. In late 1924, Fall River Iron Works Mills Nos. 6 and 7, with 100,000 spindles, 2,000 looms and 1,000 employees, moved to Kingsport, Tennessee. This marked the beginning of a fifteen-year decline in which seventy-three mills, representing three-quarters of the city's capacity, were liquidated. Sixty-three of the mills, with nearly 2.5 million spindles, belonged to just twenty-six corporations that had liquidated completely.

In 1929, eighteen mills closed, followed by nine in 1930, losing 1,221,000 spindles in just two years. In 1934, there was a second flurry of liquidations, with the Borden interests closing their remaining holdings, putting over three thousand people out of work. Ten additional mills closed in 1935, including the two remaining mills of the Durfee Corporation. The collapse of cotton manufacturing greatly affected closely associated industries, creating an unchecked whirlwind of business failings.

In response to Chief Sullivan's appeal for assistance, outside apparatus began to arrive at the fire as early as 7:45 p.m. They continued to join Fall River's forces until after midnight, adding thirty-two pumpers, ranging in capacity from 500 to 1,300 gallons per minute, as well as other equipment. Boston, fifty miles to the north, sent three engines and three hose wagons; Providence, Rhode Island, contributed four pumper engines; Brockton, Newport and Swansea sent two apparatus each; and the Naval Training Station and other communities sent one each. The smaller towns that could not send apparatus sent large volunteer crews, composed of ten to thirty men. The newspapers dubbed these volunteers the "hayseed brigade."

Fall River's fire department consisted of eight motor pumps, with a total capacity of 6,750 gallons per minute, and supplementary apparatus, such as ladder trucks, hose wagons and combination vehicles. Four years previously, Fall River and nearby cities had converted their equipment to conform to the National Standard hose coupling thread, and thus—with the exception of Newport, Rhode Island, which brought and used adaptor couplings—the outside communities were able to connect directly with Fall River's hydrants and hose. This interchangeability of hose was a vital factor in the firefighting operations as so much of the hose that was first laid immediately froze to the ice in the streets and had to be abandoned when the crews were forced to move to other locations to battle the blaze.[91]

U.S. Weather Bureau records show that during the early hours of the fire the temperature was 15 degrees and the wind was from the northwest and southwest. About midnight, the wind changed to north and northeast. The maximum wind speed was twenty-four miles per hour, and the temperature had dropped steadily to zero by six o'clock the next morning. This fast and changing wind would prove to be the fire's best friend. In several cases, the wind forced the flames to double back, creating additional destruction and confounding firefighters' efforts to get a handle on the spread of the blaze.

The late Mrs. Florence Cook Brigham, curator of the Fall River Historical Society from 1976 to 1990, was an eyewitness to the fire when she was twenty-eight years old. Her husband, Richard Curtis Brigham, was the local weather observer for the United States Weather Bureau and had called her from The Mohican Hotel, on whose roof stood the weather tower, to tell her that a fire had started at the Pocasset Mills. "He told me that one of the firemen said it was going to be 'a humdinger,'" she later remarked to reporters. "I was scared to death. I thought the whole city was doomed. It was terrible, really terrible…The only thing that kept us warm was the fire, but we had to keep moving back as it spread."[92] Coincidentally, the society she would lead in later years had its offices and exhibit room in Room 45 on the fourth floor of the Buffington Building at 10 Purchase Street. The society's entire collection was destroyed in the fire except a selection of important items that had been stored in a Herring, Hall & Marvin safe.

THE EXACT SEQUENCE OF THE SPREAD of this fire cannot be determined precisely. With Pocasset Mill No. 1 facing east, Mill No. 2 facing south and Mill No. 3 facing north, the fire fanned out in several directions about the same time. Wind carried burning embers to wooden roofs, and collapsing structures spread the flames, which made controlling the inferno impossible. Just as crews would set up in an area, the fire would take off in another direction.

While primary attention was directed toward checking the spread of the fire to the north and east, where there was a congestion of businesses, the fire swelled south across Central and Bedford Streets to ignite the Granite Block and the three big banks across from it next to City Hall. Some

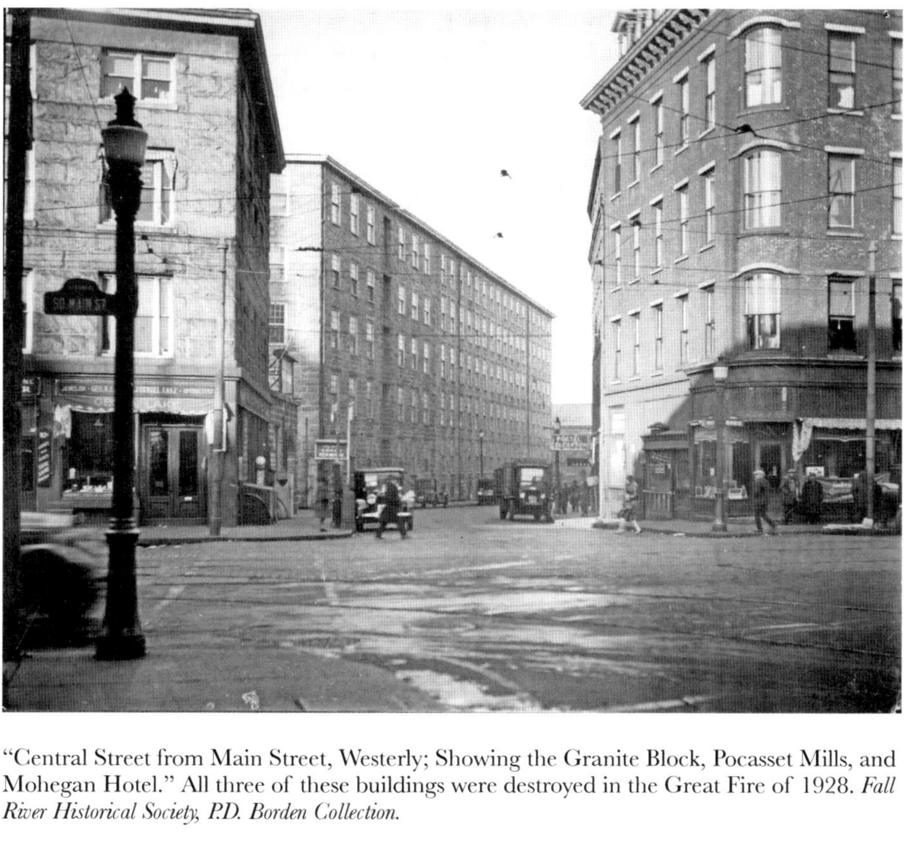

"Central Street from Main Street, Westerly; Showing the Granite Block, Pocasset Mills, and Mohegan Hotel." All three of these buildings were destroyed in the Great Fire of 1928. *Fall River Historical Society, P.D. Borden Collection.*

eyewitnesses reported that the fire crossed the street from the Durfee Block to the Granite Block (north to south) and then moved from the Granite Block to the banking establishments (west to east) by heat radiated by the wind. However, in view of the frequent changes in its direction, it is entirely possible that heat and flame were carried by the wind across Bedford Street from the Buffington Building to the banks.

When the six-story granite wall of the south side of Pocasset No. 2 collapsed at 7:40 p.m. and the flames roared across Pocasset Street and directly onto the buildings on the opposite side, three firemen were hurt, suffering injuries ranging from shock to fractured limbs. First to go was the Nathan Miller Building, followed by the old Herald Building, both made of brick. Fire lapped at the new Herald News Building next door, and the exposure to the heat from the flames was so severe that the plate glass windows cracked and shattered to the street. Because of its newer construction, however, the building was equipped with a complete automatic sprinkler system with thirty-two sprinkler heads near the windows, preventing the fire from gaining entrance. The exterior cornice burned and fell, and the building was singed; however, the interior suffered only water damage.[93]

The Mohican Hotel, part of the five-story Durfee Block, caught fire from the spread of the flames from Pocasset Mill No. 3 on Central Street. Built in 1914, The Mohican was the largest hotel in the city, able to accommodate up to five hundred guests. Unfortunately, the building contained only a partial system of automatic sprinklers, located on the street and basement levels. At about 7:45 p.m., firebrands ignited the roof, and the fire quickly spread throughout the entire building. Manager William Durfee had evacuated the guests as soon as the first fire company had arrived on the scene at the mill, and everyone was safely sent to the Hotel Mellen. Much personal property was lost in the exodus including $8,000 in jewelry belonging to Mrs. Durfee, the wife of the hotel's proprietor.[94] In less than an hour, the entire Durfee Block was burning fiercely, including the Rialto Theatre next door. Firemen directed their streams onto the roof of the building on the other side of the theater to hinder the blaze from spreading any farther north.

While the firefighters were performing their duties, there were other activities associated with a fire of this magnitude being performed. Gas company employees moved from meter to meter in advance of the fire to turn off the gas and protect structures from explosions, risking their lives in the process. From the very beginning of the blaze, the streets were filled with men rushing to save their business records from the threatened buildings.

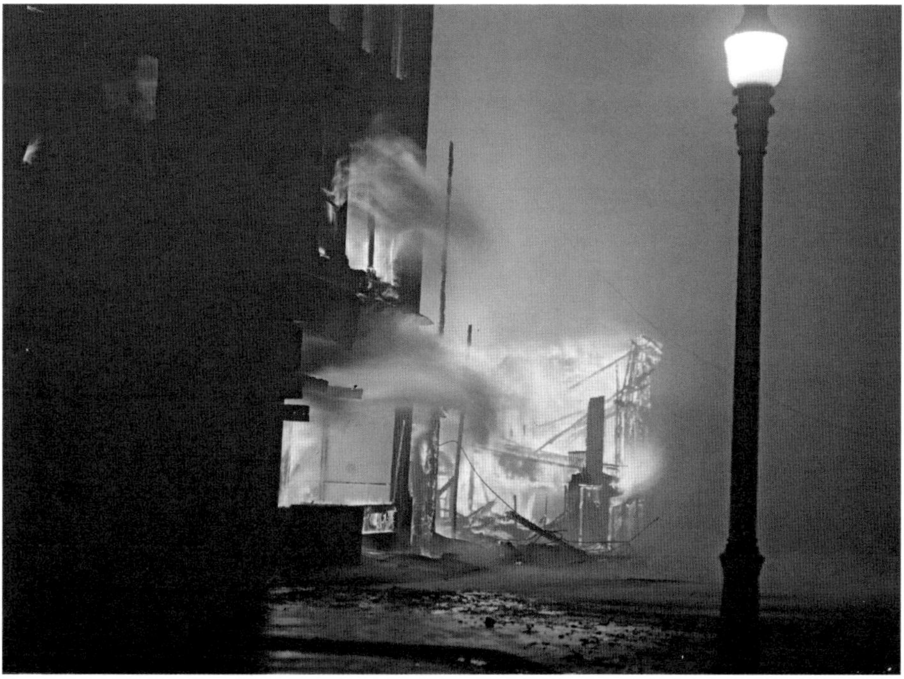

Neither brick nor wooden structures were safe from destruction during the Great Fire of 1928. *Boston Public Library, Leslie Jones Collection.*

The Metacomet Bank, for instance, had removed its entire contents prior to 11:00 p.m.

There was a bizarre story reported about the proprietor of Madam Tussaud's Museum of Celebrities in Wax, a traveling show in Fall River for a brief engagement at the Granite Block. The manager had angered a group of firemen early in the evening because he stubbornly remained in his ticket booth in case patrons wanted to view the wax figures. He was finally ordered to leave, and the firemen, none too gently, removed the wax figures to the street, where they were dropped onto the ground without thought. Several bystanders believed that these were the bodies of victims of the fire and had even called for an ambulance, which led to the rumor that there had been mass casualties downtown. The wax figures were then carried to the Metacomet Bank and lined up facing away from the street as the manager went in search of a safe place to store them. A police officer, mistaking the dummies in the smoke and darkness for people, ordered them to "move along there, move on, you can't stand here." It was not recorded what the officer said when he discovered his mistake.[95]

AT THIS POINT, the wind shifted from the northwest to the southwest, pushing the flames across North Main Street to the brick three-and-a-half-story Atwater Building on the corner of North Main and Bedford Streets. The relentless fire swept from one building to the next. The businesses bounded by Granite and Bedford Streets were no match for the flames and they quickly succumbed.

The five-story brick-and-steel Buffington Building was in the same block as the Atwater, on the northwest corner of Bedford and Purchase Streets. Said Mrs. Brigham, "Everybody kept saying that [the fire] would stop when it reached the Buffington Building because that building was fireproof."[96] This structure, unlike its neighbors, had wired glass windows in metal sashes, and these windows held. However, plate glass windows on the Purchase Street side were no defense against the severe exposure by radiated heat from buildings burning across the street. These cracked from the heat, and the fire ignited the contents on all upper floors. One office on the street level, the Aetna Life Insurance Company, was the only exception. Some of its windows were cracked, but not a single piece of furniture was damaged, even though on both sides the offices had been burned out and destroyed. Remarkably, the Buffington Building's structure was not severely damaged and was considered a good candidate for reconditioning.[97]

To the north of the Buffington Building, in the next block, was a gas station on the corner of Granite and Purchase Streets. Around this business was a sizeable amount of open space, and it was thought that the easterly spread of the fire might be checked at this location. However, the same change in the direction of the wind that had checked the fire in its northerly spread at Bank Street worked against the fire being stopped here. Amazingly, it was later determined that no gasoline, fuel oil or city gas had contributed in any way to the brutality of this fire. The gas tanks here were underground, and the small amount of gas in the pumps was "quietly burned off" without incident.[98]

At about 8:30 p.m., the flames from the burning Atwater Building on Bedford and Main Streets spread to the four-story brick New Wilbur Hotel located at the corner of Granite and North Main Streets. Renovated in only the past year, this structure could not survive against the intensity of the inferno fueled by the high winds. The next building north contained the offices of the *Fall River Daily Globe*. According to the coverage that paper published the next day, employees of the Wilbur and the *Globe* halted the progress of the fire for a while at the hotel, but the deadly combination of wind and fire won out when the roof ignited and the flames swiftly traveled from floor to floor until the entire hotel was consumed.

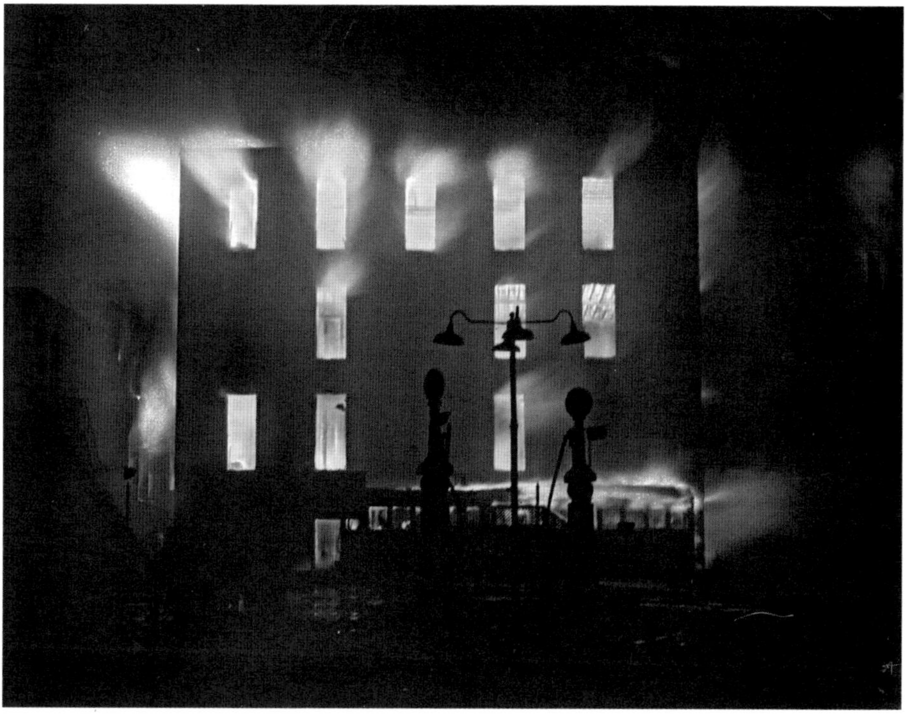

The east side of Printers' Hall, ablaze, on Granite Street, with the Smith Oil Company filling station in the foreground. Amazingly, this gas station, located on the corner of Granite and Purchase Streets, was not ignited in the Great Fire of 1928. *Boston Public Library, Leslie Jones Collection.*

The northerly spread of the flames next attacked the Temple Beth-El Synagogue that faced Bank Street and extended back into the center of the New Wilbur Hotel block. Principally of wooden construction, the synagogue was totally destroyed but not before Rabbi Morton Goldberg had rushed into the burning temple to save the holy scrolls, the canopy and some valuable religious texts. Unfortunately, the rabbi lost his priceless personal library that he had been collecting for fifteen years.

To the east of the temple sat the two-story brick New England Telephone Company. Wooden tin-clad shutters and wired glass windows, reinforced by outside sprinklers, prevented the entrance of fire to this building, but the smoke was so thick that it was necessary for the operators to abandon their stations for a period of six hours, the structure being evacuated about 11:00 p.m. Two fire crews from Brockton saved the telephone exchange building, but all the phone lines were out of order from the water that was poured onto it. With the failure of telephone service, all communication

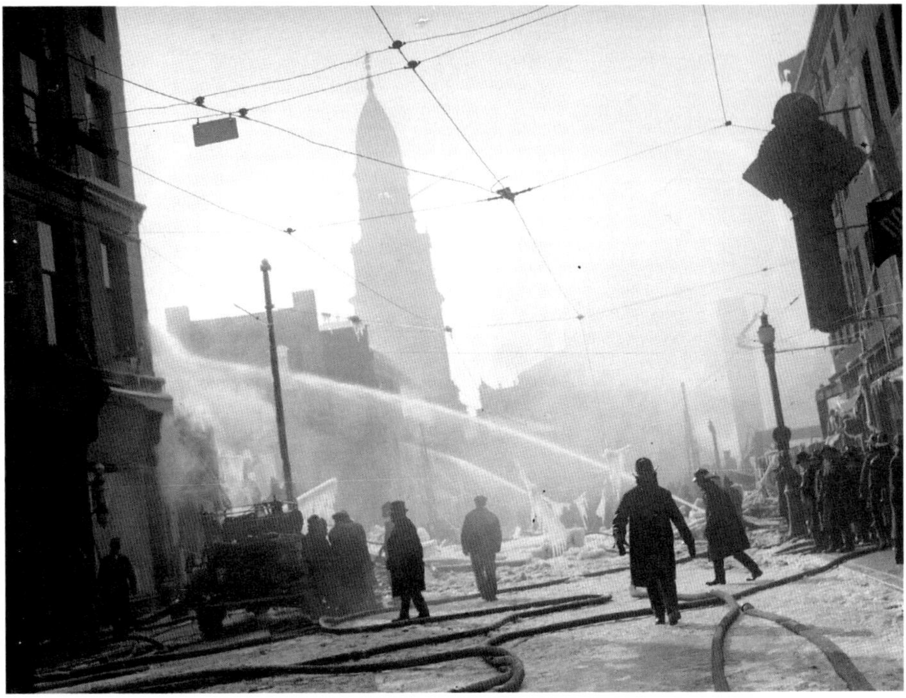

The Great Fire of 1928 still smoldered on the morning of February 3 as firefighters continue to hold it back at the Fall River Daily Globe Building. The ruins of the New Wilbur Hotel were sprayed with multiple streams of water. In the smoky background, you can see the tower of city hall, untouched by this blaze. *Boston Public Library, Leslie Jones Collection.*

between Fall River and the outside world was essentially cut off—both the Postal and Western Union Telegraph offices had been destroyed early in the conflagration.

In dramatic fashion, the fire was finally checked in its northerly spread along North Main Street by the firewall and open sprinklers protecting the south side of the Globe Building and the few windows that opened into a courtyard on its east. These open sprinklers operated so successfully that the ordinary thin glass in these windows was not even cracked. And while the fire did not advance any farther north, other even more massive catastrophes awaited.

Witnesses reported that the east side of Main Street was an inferno, as building after building caught fire, defying all human attempts to control the flames furiously burning within them. As the night wore on, the wind increased in velocity, and it seemed that there would be no end to it.

By 9:00 p.m., the blaze had spread to cover Fall River's entire central downtown business district. Pumper trucks were strategically placed along

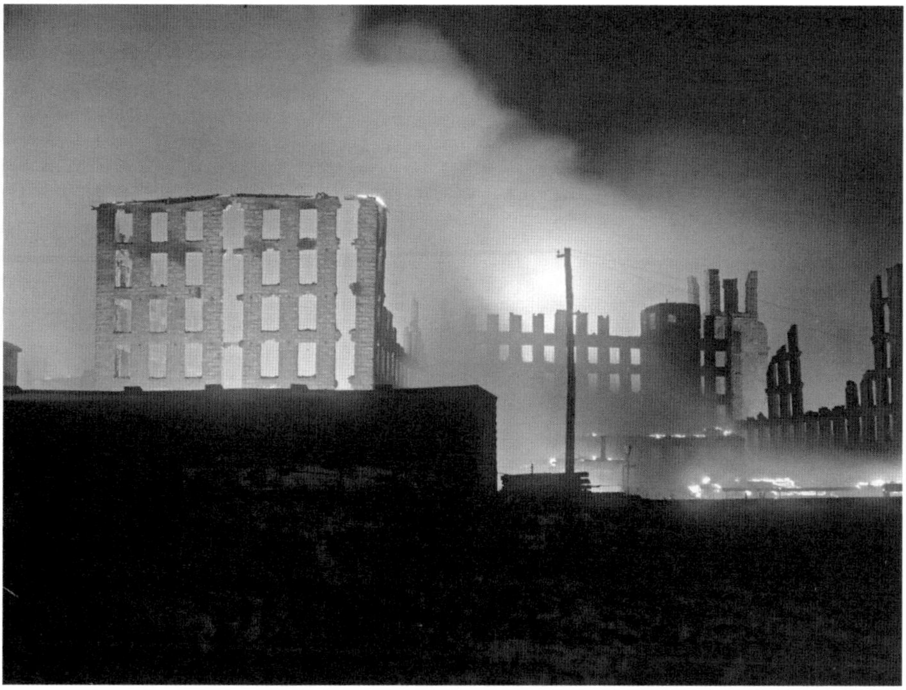

The burning of the Pocasset Mills around 2:00 a.m. on the morning of February 3. *Boston Public Library, Leslie Jones Collection.*

the street as more out-of-town help arrived throughout the night. Try as they might, the fire companies could not contain the area of the conflagration. The blaze on the east side of Main Street spread from the Waring Building to the Premiere Theatre on the corner of Granite and Rock Streets, a two-story wooden structure that was little more than fuel to this fire.

The Metacomet National Bank, at the corner of Bedford and Rock Streets, and the string of buildings bordering Bedford Street as far as the police station were also ignited from flying sparks. This bank had recently undergone an extensive and expensive renovation; however, the work was cosmetic only. No firewall had been added to the structure between the bank and the other buildings to the west, so it was relatively easy for the fire to enter through unprotected openings and practically destroy this structure. The continued spread of the fire was finally checked in this direction by considerable efforts on the part of firefighters. The Great Fire of 1928 was still just getting started.

The Interstate Bus Terminal, located behind the Granite Block, was destroyed when the front-facing wall from Mill No. 1 fell directly onto it.

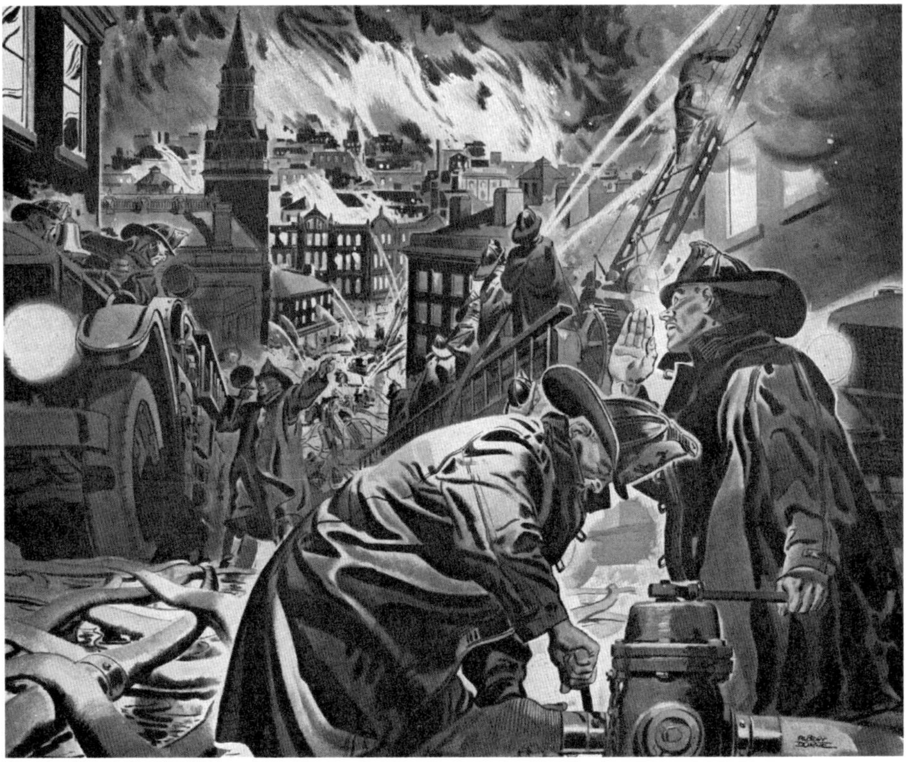

A City Saved by a Thread depicts fighting the Great Fire of 1928. *Magazine advertisement for Capital Stock Fire Insurance.*

The northern end of the landmark eighty-four-year-old Granite Block was observed to be on fire about 9:00 p.m. In this four-story granite structure, sixteen businesses were located on the street level, and sixty-five others were on the upper floors. The Granite Block had two firewalls across the building, but neither one reached the roof. So instead of stopping the fire, they merely slowed it down. The fire spread through the structure from north to south, and it was four hours later, close to 1:00 a.m., before the destruction was complete.

Sadly, it was reported that in the weeks before the fire, plans had been under way for the installation of a complete system of automatic sprinkler protection in the Granite Block. If this improvement had been completed prior to the fire, it was believed that, in all probability, this building would not have been destroyed.

While the Granite Block was burning, Engine 26 from Boston protected City Hall and the post office, both fire-resistive granite structures, though without the window protections, such as wired glass, metal sashes, tin fire shutters or open sprinkler systems, that had saved other buildings this night. The firemen effectively prevented the fire from entering the windows of these buildings by covering them with water curtains from hose streams. According to the *Boston Globe*'s reporting:

> *This crew stretched the first line of 500 feet from Engine 4 of Providence into a ladder pipe mounted on Ladder 4 of the Fall River department, and made it possible to operate this pipe as a water current between a big seven-story office building, which was burning from cellar to roof, and the City Hall and Post Office. The operating of this stream prevented the flames from getting a hold on the other two buildings, and gave the firemen from half a dozen different cities working along Second Street a chance to stop the fire at that point.*[99]

By 1928, the automobile had become a popular mode of transportation—so much so that when the news spread of the fire in Fall River, hundreds of thousands of people made their way to the city to watch the event firsthand. All roads leading into the city were jammed with vehicles of all types, inching their way to the bright, glowing sky in front of them. Every so often, a fire apparatus and its crew would approach and the long line would part to allow them passage. It was reported that an astounding 250,000 sightseers made their way to Fall River that night and in the days that followed, prompting Chief of Police Martin Feeney to call for militia to aid in patrolling the fire zone, in which martial law had been declared.

The first call for militia was sounded at 12:10 a.m. and the second call at 1:25 a.m. A large detachment of state police was sent to the scene as well as four companies of militia, naval reservists and members of the Coast Guard. The troops were from the Third Battalion of the 241st Regiment. One hundred men were also sent from Newport Naval Training Station.[100]

A funny story was reported in the *Boston Globe* with the headline "Guardsmen at Fall River Obey Rules Too Literally" that deserves retelling:

> *The best story of the whole fire, however, concerns Judge Frank M. Silvia, Justice of the 2nd District Court, and* [Fall River's] *Mayor* [Harry] *Monks. Judge Silvia, attempting to go to his chambers in the courthouse, in Rock St., was ordered back by one of the National Guardsmen. Explanations made by the judge did no good. The youth was adamant. He had his orders and he was going to obey them.*
>
> *The judge turned away and went looking for Mayor Monks. He explained his predicament to the chief executive, who said, "O, that's all*

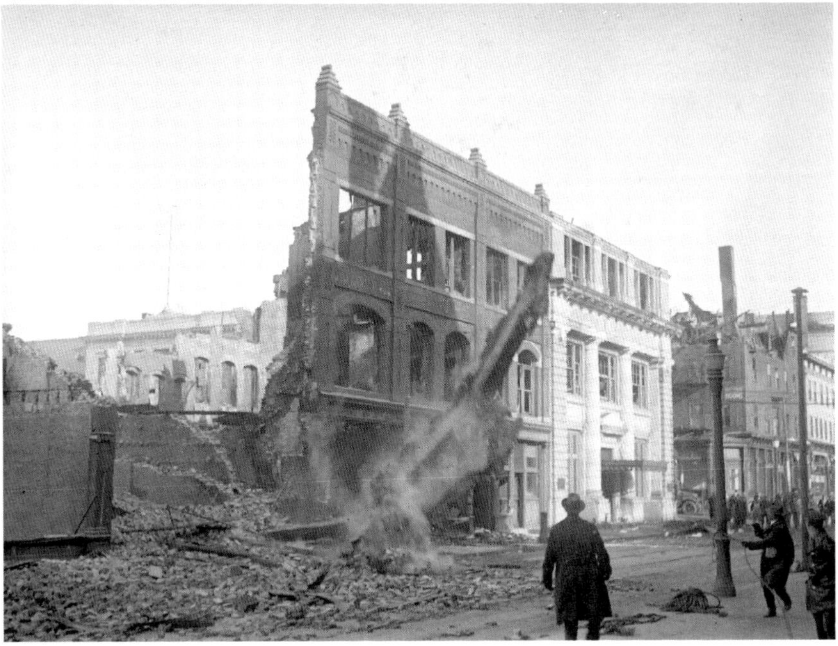

"Dynamiting Fall River fire ruins," 1928. *Boston Public Library, Leslie Jones Collection.*

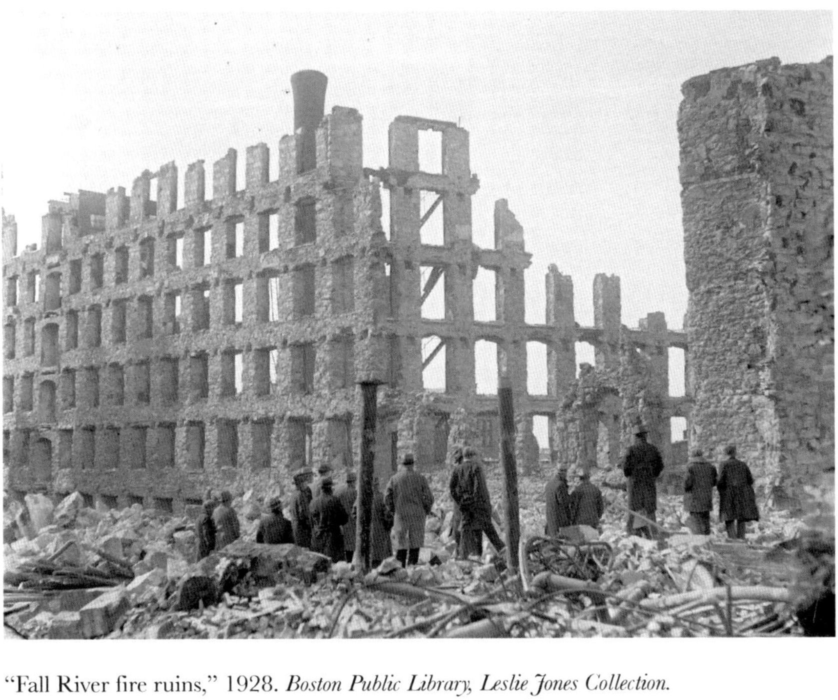

"Fall River fire ruins," 1928. *Boston Public Library, Leslie Jones Collection.*

right, I'll get you by. Incidentally, I think I'll drop into the Courthouse with you for a little while. I've got some things I want to talk about."

The two walked back to Rock St., where the young man came smartly to "port arms."

"You can't get by here," he said.

Mayor Monks made a conciliatory gesture. "Now, listen," he said. "You've to use reason in all things. This is Judge Silvia and he wants to get into the Courthouse. Surely he ought to get by."

The young guardsman thought it over for a few minutes, then decided to relent.

"All right," he said, "that sounds reasonable. Go ahead, Judge."

Judge Silvia started past the deadline and Mayor Monks fell in step.

"Just a minute," cried the youthful soldier to the Mayor. "Where do you think you're going?"

"I'm going to the Courthouse with the Judge," said His Honor.

"O, no, you're not," said the youth.

"Young man," said Mayor Monks sternly, "do you know who I am? I am the Mayor of this city."

"I know that, all right, but"—and the young main pointed toward the southeast—"City Hall's over there."

The block next to City Hall on the north and bounded by Bedford, Second, Main and Market Streets was occupied by three of Fall River's largest banks: Citizens Savings on the northwest corner, the Union Savings on the southwest corner and the Massasoit-Pocasset on the east. It is unclear exactly when the Citizens Savings caught fire, but around 2:00 a.m., the Massasoit-Pocasset abruptly burst into flames. Even though the Union Savings Bank was reported to be of fire-resistive construction, it had a plank-and-timber roof supported on a steel frame. Firefighters believed that the building was saved by the several inches of water they had sprayed onto the roof from their hose streams. But according to witnesses, the fire snuck in through a plain glass skylight, and this building was doomed to go the way of the neighboring banks.

The strong winds carried embers and firebrands like snowflakes beyond the burning district to other parts of the city. The Troy Cotton and Woolen Manufactory was about seventy-five feet away from the downtown conflagration, but its stairway tower ignited just the same. Workers there attached fire hoses to yard hydrants, but the pressure proved to be weak, prompting the mill engineer to start up the one-thousand-gallon fire pump.

Hose streams were carried onto the roof, which was kept constantly wet, and the fire in the tower was extinguished with only minor damage.

Nearly half a mile east of this mill, the firebrands ignited the tall tower on Mill No. 1 of the Merchants Manufacturing Company. Here again, it was necessary to rely on the mill's pumps for sufficient water pressure. Fire apparatus was dispatched from the downtown conflagration and carried hose lines to the roof in order to reach the flames. The fire was

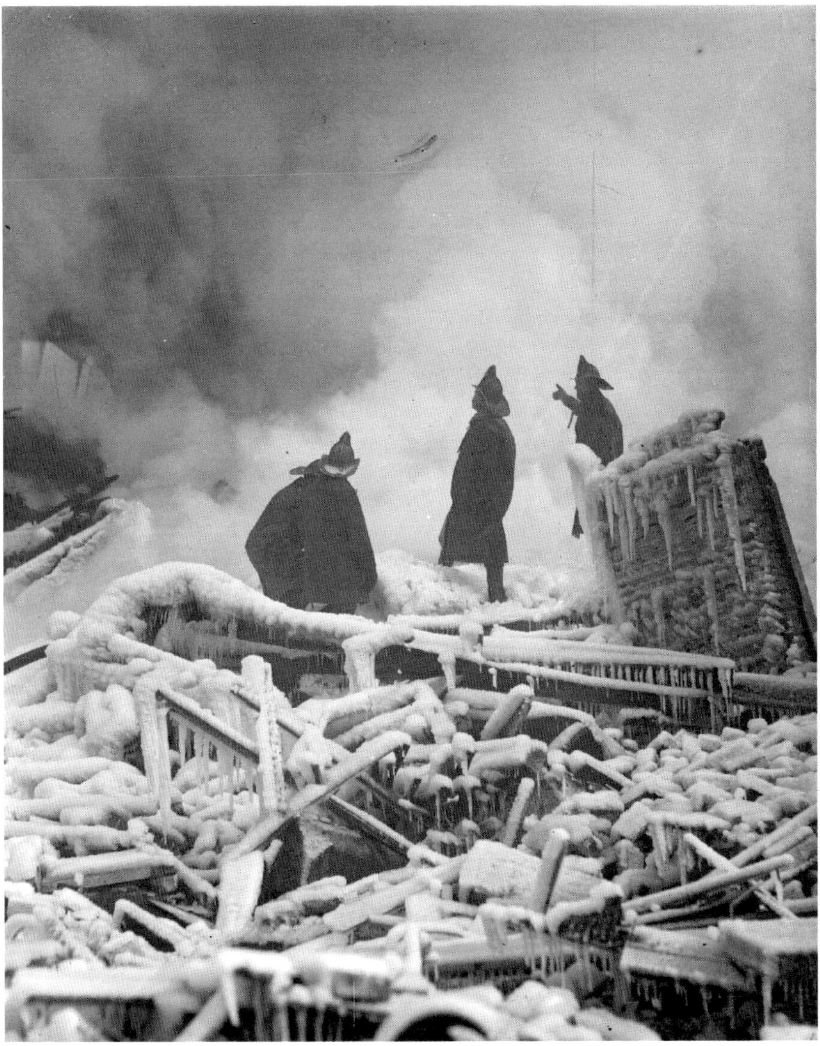

Four Fall River firefighters survey the damage from atop an ice-covered debris field following the Great Fire of 1928. *Boston Public Library, Leslie Jones Collection.*

finally extinguished but "not before the roof of the tower had been entirely destroyed and considerable water had run down the tower and spread into the mill building, wetting some machinery and stock." The mill had employed waterproof covers for its equipment and goods, so the total damage was only about $5,000. "Had it not been for the good work with the ample pressure maintained by the pumps, the old mansard roof of this would surely have been ignited and a large loss would have followed."[101]

Fall River had enacted an ordinance in 1915, before the Steiger Store fire, requiring that all new construction use fire-retardant roofing materials. But even thirteen years later in 1928, many of the old wooden shingle roofs remained, and because of that fact, some twenty-five smaller fires were started by burning embers that landed on roofs like these as far away as three-quarters of a mile from the center of the main fire.

Chief Jeremiah Sullivan declared the Great Fire of 1928 under control at 2:30 a.m., even though it would be four days before the last apparatus was recalled from the devastated and still smoking ruins.

Aftermath

Throughout the fire area, practically all of the contents contained in wooden and steel filing cases were destroyed. Those business owners who had not risked life and limb to salvage their account books and legal document learned in the morning that the fire had taken their files, records and fiscal accounts—an untold and irreplaceable collection of important papers. The offices of the Income Tax Bureau had been in the Buffington Building, and four thousand state income tax returns filed since the first of the year in Bristol, Barnstable, Dukes, Nantucket and Plymouth Counties were now just so much ash. It was announced that anyone who had already paid their taxes had to re-file and include a statement to this effect on the new form.

The loss of files and paperwork proved to be a stroke of good luck for a Brooklyn woman, Mrs. Catherine Barletto, twenty-six, formerly of Fall River. She was to be the first woman in New York to be sentenced to life in prison under new crime laws for her fourth robbery conviction, the first three of which occurred in Fall River. Because her records were burned in the fire, her first three convictions could not be proved, and it was estimated that the most she would serve would be five years.[102]

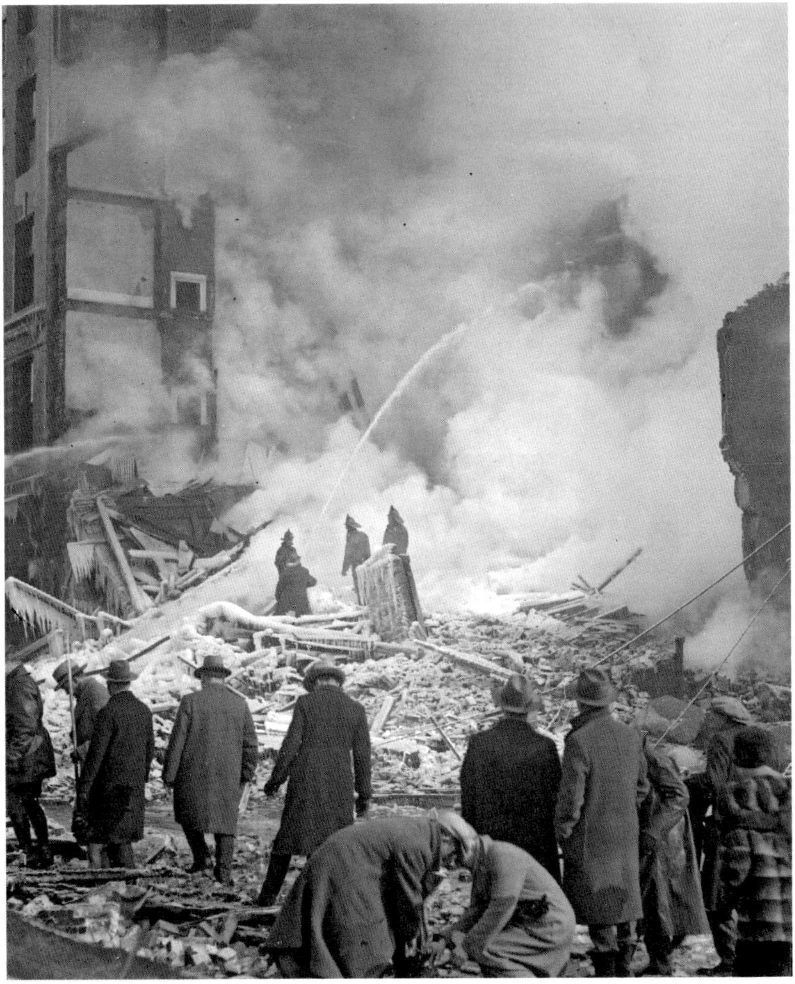

Officials survey the devastation in downtown Fall River as firefighters continue to stream water onto the smoldering ruins. *Boston Public Library, Leslie Jones Collection.*

In addition to the more open method of storing documents, it was conservatively estimated that there were 125 safes of all types involved in this conflagration. These businesses fared better as most of the safes with a high fire-resistance rating worked properly and preserved their contents. Only a few safes that had been severely exposed to fire and buried in the smoldering ruins of the Granite and Durfee Blocks fared poorly. In two cases, it was reported that safes had been blown open by "the internal pressure of gasses formed by the rapid heating of the insulating material or contents."

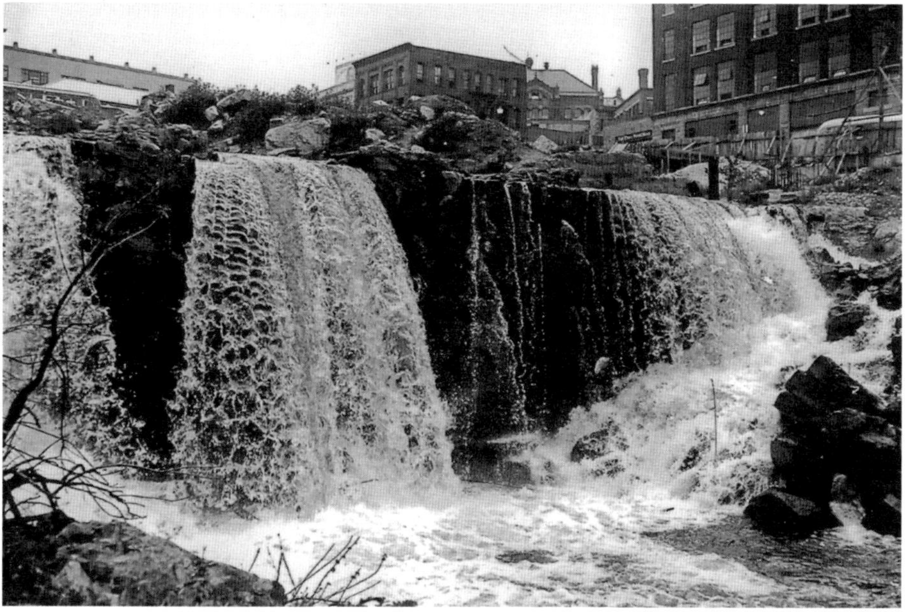

The falls were exposed for the first time in eighty years when the Pocasset Manufacturing Company burned down in the Great Fire of 1928. *Fall River Historical Society, P.D. Borden Collection.*

Generally, safes with only two or three inches of insulation did not survive, although in a few instances, they did, "presumably having been spared any severe exposure owing to some freak of the fire."[103]

Mrs. Brigham said that her father, Judge Benjamin Cook Jr., was optimistic that his legal papers stored in his office would survive the inferno because they were in a fireproof safe. "What he hadn't figured on was that the safe would fall into the Quequechan River and that its door would be knocked off by a granite ledge. All the contents were lost."[104]

Of the seven banks burned in this fire, four contained the typical bank vault—massive walls and heavy steel doors—which preserved their contents perfectly, even exhibiting no signs of water damage. The exception to this statistic was the basement storage vault of the Union Savings Bank. "This particular vault was completely submerged for several hours and became filled with water. Water rose in the book vault above it to a depth of six inches and the contents absorbed water to a height of two feet above the floor."[105]

When the sun rose on the morning of the third of February, the 150-foot chimney of the Pocasset Mill offered a strange sight, "standing solitary amidst the ruins of the mill." To the amazement of onlookers,

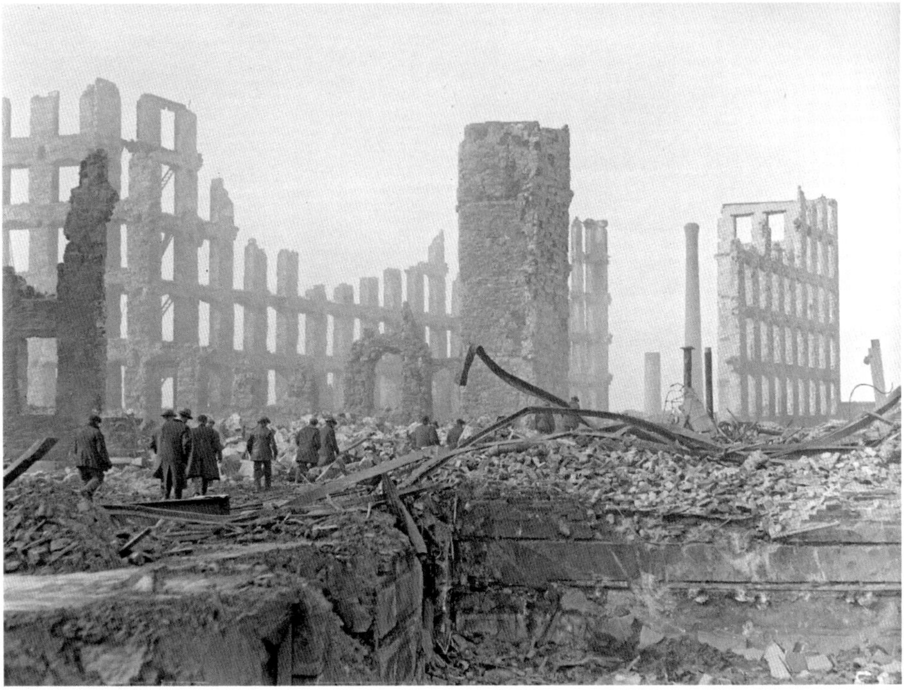

"Pocasset Mills fire ruins, Fall River," 1928. *Boston Public Library, Leslie Jones Collection.*

smoke was seen pouring from the chimney for the first time in two years as the updraft sucked the fumes from the smoldering ruins and released them into the air.

With the exception of a part of the south end, which was demolished several days later, all of the walls of the Granite Block collapsed during the fire. Likewise, the remaining walls of the Pocasset Mills either dropped during the fire or were taken down during the city's massive cleanup efforts. "Brick walls in the burned area fared little better, there being very few walls retaining sufficient integrity to be considered safe for use in reconstruction."[106]

Dynamite teams arrived in the city the next day, and after much study and engineering, the five-story, 210-foot-long north wall of the Pocasset Mill No. 3 was brought down ten days later. This wall was the most dangerous of the remaining structures because it was standing unsupported by frame or timber, except for its own weight. When it fell, stones flew in all directions from the explosion and subsequent collapse.

On North Main Street, the Globe Building, whose firewall just ten days before had stopped the northern progression of the conflagration, was

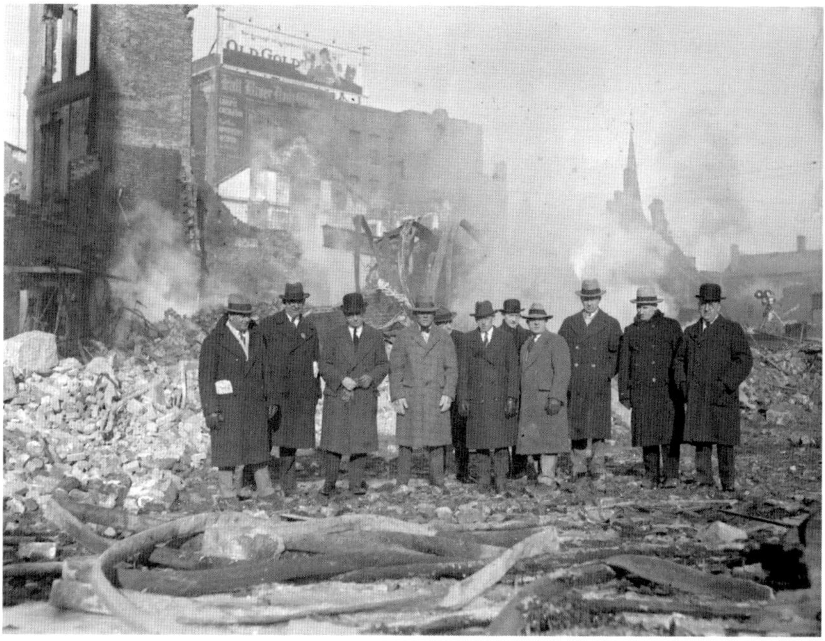

"Fall River fire, dynamiting ruins," 1928. *Boston Public Library, Leslie Jones Collection.*

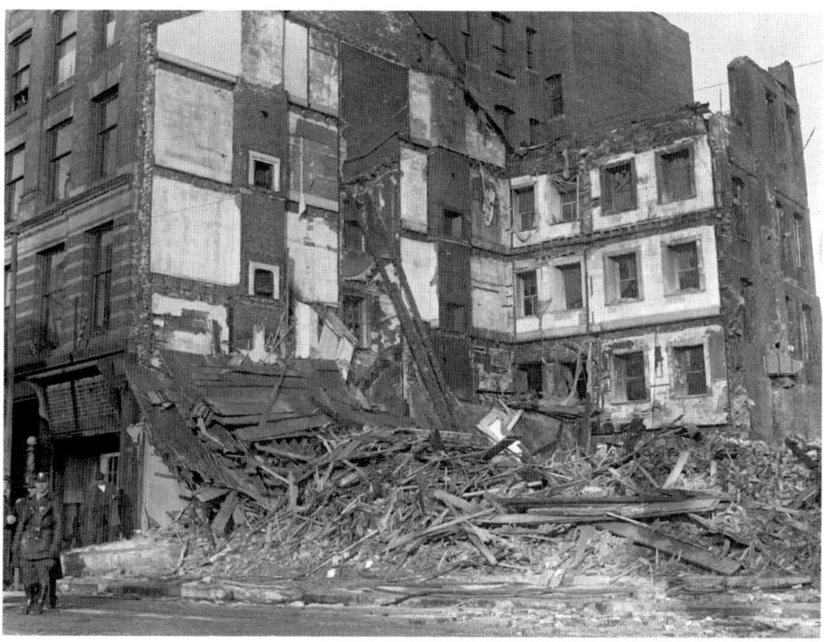

"Where they stopped the big Fall River fire at the Fall River Globe building on Main Street," 1928. *Boston Public Library, Leslie Jones Collection.*

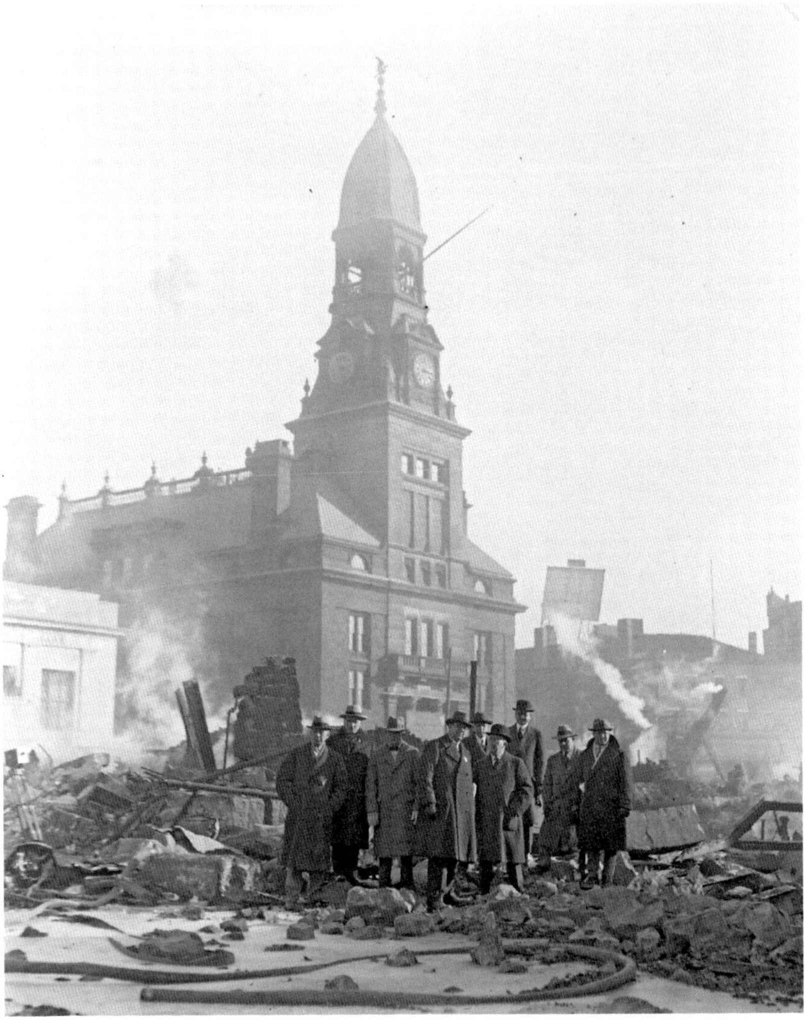

"Fall River fire: City Hall and dynamite experts looking over dangerous Pocasset Mills walls," 1928. *Boston Public Library, Leslie Jones Collection.*

hit by a cantaloupe-sized chunk of granite, smashing a window of the proofing room on the fifth floor. Some of the flying stones also reached the Masonic Temple on North Main Street, three blocks away. Several innocent bystanders were injured but none seriously.

And within an amazingly short eighteen months, all of this destruction and ruin had been rebuilt. Contractors announced that more than 350 men would be hired as soon as the foundations of the proposed structures were

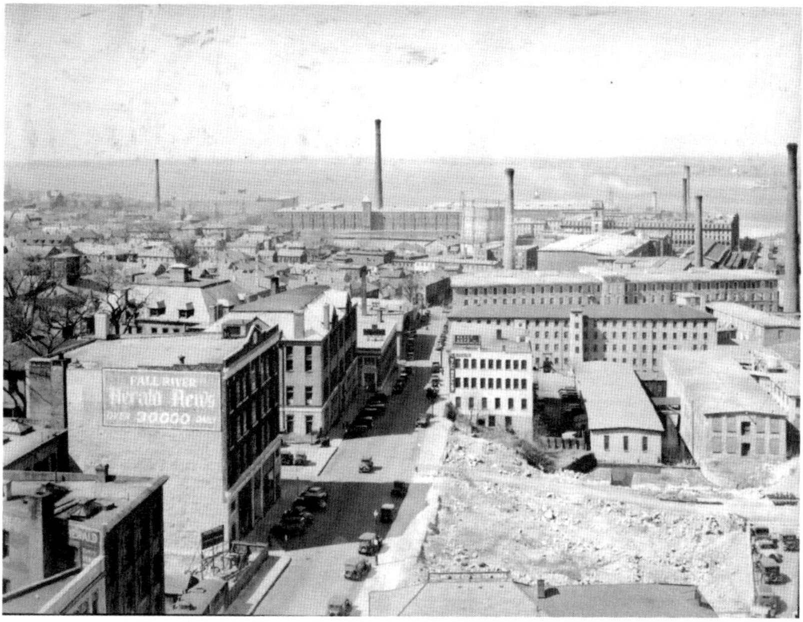

"Looking down Pocasset from City Hall, 1928." *Fall River Historical Society, P.D. Borden Collection.*

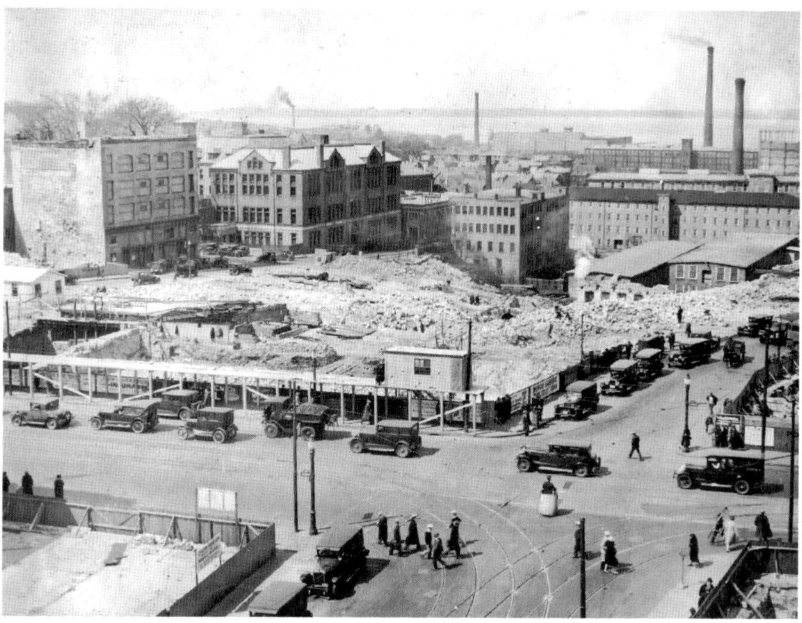

"Looking at the ruins of the Granite Block and Pocasset Mill from City Hall, 1928." *Fall River Historical Society, P.D. Borden Collection.*

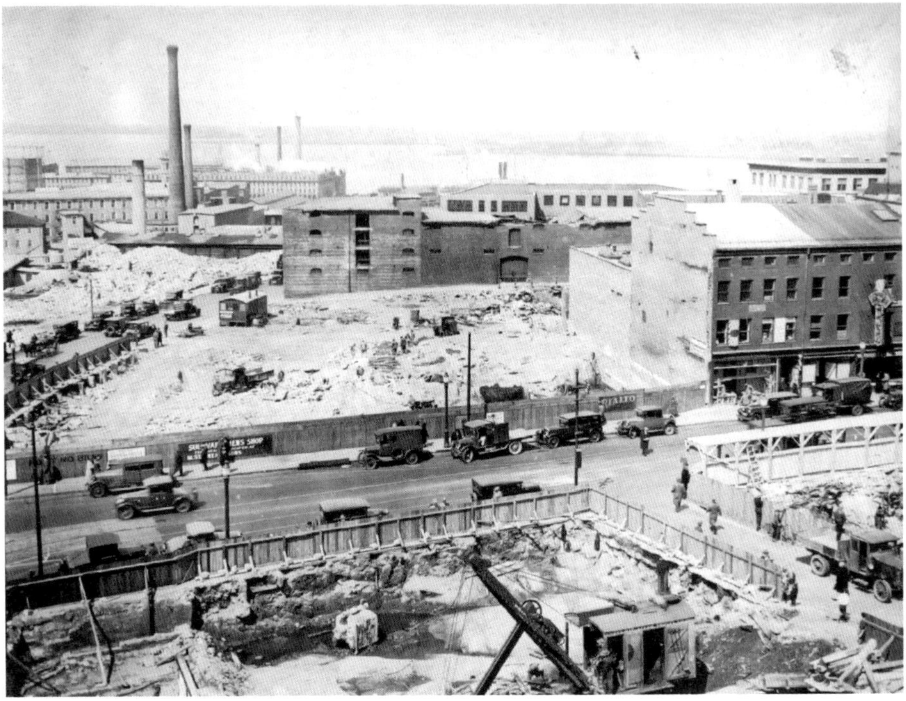

"Looking at the ruins of the Mohican Hotel from City Hall, 1928." *Fall River Historical Society, P.D. Borden Collection.*

begun, with day and night shifts, on schedule to be completed by October 1, 1929.

Ground was broken on May 28 for a new Temple Beth-El on High and Locust Streets, the front of the Herald News Building was rebuilt, the Buffington Building was rehabilitated and the devastated area was modernized with wider streets that could accommodate automobile traffic and pedestrian use, could serve as reasonable fire stops during any future conflagrations and would meet the demands of new firefighting apparatus. But perhaps the biggest symbol of progress following the fire was the construction of a new Granite Block, at a cost of $750,000. Just like the rebuilding frenzy that had occurred following the Great Fire of 1843, Fall River was determined to show its residents, business owners and the outside world that it could be depended on to rise to the occasion and come back from any adversity.

As a symbol of this spirit of optimism, Mayor Monks announced that the city would "build a park out of the ashes." The fire debris was taken

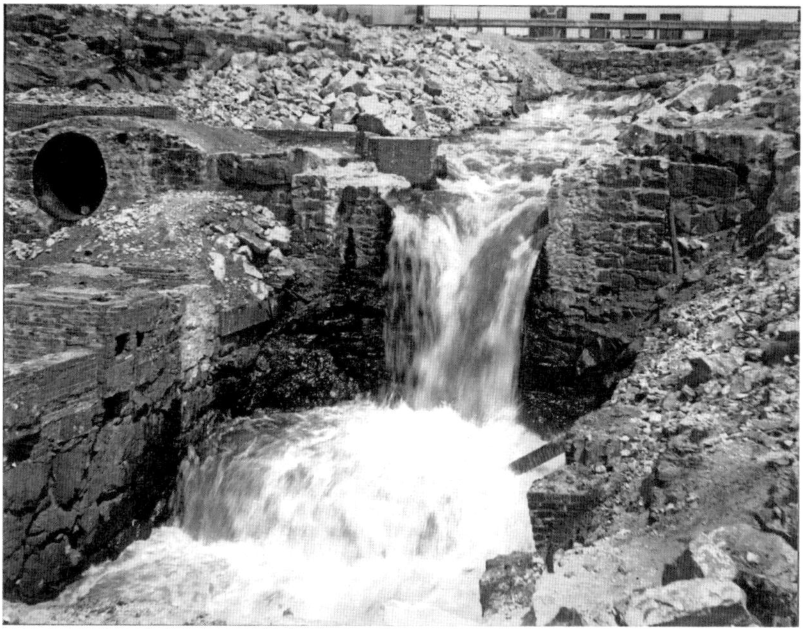

"First Falls in the River under the former Pocasset Manufacturing Company, in 1929." The rubble from the destruction of the Pocasset Mill is clearly seen here. *Fall River Historical Society, P.D. Borden Collection.*

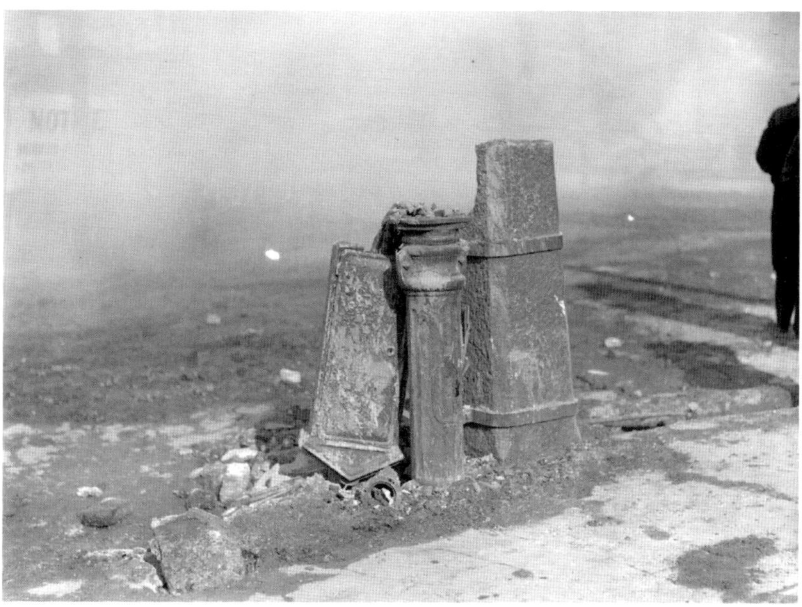

"The alarm box that was rung during the Fall River fire. The alarm was heard around the country," 1928. *Boston Public Library, Leslie Jones Collection.*

to the Globe Pond located at the corner of Globe and South Main Streets, where the first cotton mill was established in 1811, now long gone, and filled in to make a park. And so Father Kelly Park was created and exists today, even though residents are probably quite unaware of the significance of this location and the history that is literally under their feet.

Billboards were soon raised that proclaimed the indomitable spirit of Fall River following the Great Fire of 1928: "Yes, we had a fire. It was a real one. But we also have a big city with real people in it. We are going to make it bigger and better than ever."

5

THE NOTRE DAME DE LOURDES FIRE OF 1982

It took one-half hour to level a church that took sixteen years to build.
—Fire Chief Louis Shea

Of all of Fall River's horrific conflagrations, the fiery destruction of the Notre Dame de Lourdes church and surrounding residences and businesses was perhaps the most emotional loss that the city has ever endured. To this day, those who witnessed it or were affected by it remember, in vivid detail, the 235-foot copper-sheathed twin spires collapsing, the 100-foot-wide and 235-foot-long structure crumbling and the emergency evacuation of seventy-five neighboring families. To many it was like a death in the family. Even the normally stoic *Fall River Herald News* led with this headline the morning after: "The Day a City Wept."

Yet the story of this fire is much more than just the actual event itself. It is a tale full of history, cultural pride and a stubborn insistence for a spiritual leader who was one of them. For the chronicle of the building of Notre Dame is just as fascinating as the story of the blaze that brought it down.

In the 1870s and '80s, Fall River's textile industry was in an aggressive era of expansion. At the same time, the mainly agricultural residents of Lower Quebec Province in Canada were experiencing a harrowing life of poverty, overpopulation and debt as they attempted to eke out a living in a largely infertile environment. With New England close by and a surfeit of unskilled or semi-skilled jobs available in the burgeoning cotton textile manufacturing trade, French Canadians flooded into Massachusetts. By

1900, they would "convert Fall River into the largest French-Canadian community in New England in terms of total population."[107]

The French Canadians, for the most part, settled in the Flint section of Fall River, named for John D. Flint, who had arrived in the village in 1847 and eventually had two farms that covered the eastern area of the city. Both this neighborhood and the Flint Mill, built in 1872, were named in his honor.

As a cultural minority, the French Canadians felt the intense need to preserve their ethnic identity, including their language and attitudes about labor, politics and religion, as a kind of protection against those disturbed by their mass influx. Organized labor believed that this group was, as a whole, much too submissive to management, accepted lower wages for their work (which affected the wages of all workers), violated child and hour law regulations, was opposed to strikes and was uninterested in trade unionism. In fact, in 1879 their pastor and community leader, Reverend Pierre Jean Baptiste Bedard, spoke out against walkouts, encouraging his congregation to stay out of the fray, even going so far as to communicate with Canada to procure strikebreakers. Bedard knew that the French Canadians of his parish hadn't been in Fall River long enough to compile savings or reserves as a cushion against unemployment. It was his belief that if striking workers had the right to leave work for their reasons, then they must respect the right of others and not interfere when the needy accepted positions offered to them. Obviously, these actions did not engender trust or acceptance among the native-born population or assimilated immigrants of the city.

The first French Canadian church in Fall River was St. Anne's, located on South Main Street and dedicated in 1870. In 1872, the Catholic churches in Fall River became part of the newly created diocese of Providence, administered by Bishop Thomas F. Hendricken, an Irishman who believed in a doctrine of assimilation. He strived to impress on newly formed parishes the need to shed cultural identities in favor of a homogenous religious character that would help ease the fear Protestants of the area had for Catholicism as a foreign religion and make administration of the diocese easier. If the native populations could see that Catholics were just like them, there would be far greater acceptance and thus the possibility of converts to the faith.

The parish of Notre Dame was founded in the Flint section of Fall River in 1874. The first pastor was the aforementioned Reverend P.J.B. Bedard, a thirty-two-year-old French Canadian known for his powerful voice and large physique who had been a vicar in several rural Canadian parishes. The first Mass held at Flint village was before a crude altar of planks and lime barrels at a house then being built for Irishman Henry McGee, a grocer. At the time, a subscription list was taken and within six weeks a wooden chapel had been constructed. By 1877, a school and a convent had been built, run by the newly arrived Sisters of Jesus-Marie, of Sillery, Quebec.

Bedard was an energetic and charismatic leader, attracting great crowds to his sermons. He ran his parish contrary to the assimilation policy of Bishop Hendricken, encouraging his flock to keep the national spirit, religion and culture of Canada alive. "The French-Canadian immigrants were bound by code of 'survivance,' a sense of the need to preserve their language and culture in religious practice."[108] This aggressive nationalism had estranged Bishop Hendricken from him and would prove to have repercussions long after Bedard's sudden death in 1884 from "apoplexy."

Bedard had purchased land for the bishop, including the parcel where the Notre Dame de Lourdes was built and some adjoining plots. With an inheritance of about $3,000, he bought large tracts adjoining this area and then sold a large portion when real estate was at its peak, profiting some 5,200 percent. He devoted a portion of his money to building the Jesus-Marie convent, which was originally an orphanage. He paid for the high school and purchased a home that was converted into a rectory—all with his own funds rather than those of the parish—paying municipal taxes on Notre Dame's educational and charitable buildings. The city's tax assessor suggested that by incorporating the property as a religious and charitable organization, he would be tax exempt, according to state law.

In April 1883, Reverend Bedard became president of Notre Dame Corporation and appointed the Sisters of Jesus-Marie as his assistants. In 1884, he relinquished control to a group of nine male parishioners and, at the same time, awarded the corporation his entire personal property, to be sold as the corporation willed. The parish additionally earned about $5,000 per year from the sale of granite quarried on a section of this land. The structuring of this corporation and its holdings would prove to be an additional wedge that would come between Bishop Hendricken and the Notre Dame parish. All the pieces were now in place for what was to be infamously referred to ever after as "the Flint affair."

When Bedard suddenly died in 1884, Hendricken thought he would finally be able to exert his concept of a territorial over an ethnic parish and create an English-language diocese. An interim successor was appointed, E.E. Nobert, a French Canadian priest; however, he was not well, and the parish was too large for him to manage. So he requested another assignment. Then Hendricken appointed Reverend Samuel P. McGee, whose brother was the grocer in whose store the first Mass was held. McGee was Canadian-born but of Irish parentage and, as such, unacceptable to the parish elders. The corporation laymen asked the bishop to reverse his decision, against which Hendricken railed, noting their excessive ethnic pride and claiming that McGee could speak better French than they. In retaliation against this perceived insult and the bishop's refusal to acquiesce, the parish members boycotted McGee and offered him no more baptisms or marriages, usually a source of priestly income.

The way the corporation was structured meant that Hendricken could appoint a successor to Bedard, but the laymen could order the priest from any corporation property. In late 1884, the corporation laymen collected offerings at Mass but refused to turn over to McGee his portion. Additionally, they gave McGee two weeks to vacate the rectory, which he did, moving into his brother's house on Pleasant Street. One night, whiskey bottles filled with sewage where thrown through a window at McGee as he slept, exploding on him and filling the room with the filth. Fearing for his life, McGee went to the mayor of Fall River and asked for police protection.

In 1885, a group of fifty parishioners restrained McGee by the rear altar, and one Sunday, they blocked the entrance to the church. It got so bad that when McGee would make announcements following the Mass, people would walk out. When he changed the order of things and read them first, the parishioners coughed and sneezed until he was done. The parish then voted to send a delegation to Rome to appeal Hendricken's decision to appoint McGee and not a French Canadian.

In other acts of religious disobedience, the choir of Notre Dame would not cooperate, so McGee organized a new choir and on the date of its first performance was unable to get into the choir loft, as the old choir had locked the entrance. The old choir refused to obey his command to leave, so the pastor summoned the police. There were four arrests, and the judge assigned a one-dollar fine to each. One Sunday, Irish Catholics from Immaculate Conception parish gathered in support of McGee, prompting the French Canadians to ask the mayor for police protection as well.

McGee was finally removed from the parish by the bishop, who promoted the curate from St. Anne's, Owen Clarke, as pastor, assigning Father Briscoe as the new Irish curate in that parish. When Notre Dame's parishioners boycotted Mass in protest for this choice, Clarke fled to Immaculate Conception and asked the bishop to relieve him. This ongoing battle of wills reached a tipping point when Hendricken placed the parish under an interdict, the harshest of ecclesiastical penalties. Clarke was instructed to remove all holy objects from the church and announced that all care for the sick and those seeking spiritual guidance would be administered at all other Fall River Catholic churches; such consideration was specifically denied to Notre Dame parish leaders as a penalty for causing the interdict.

In response to this act, the parishioners voted to hire two canon lawyers to present their case for a French Canadian curate, of French Canadian descent, to the Sacred Congregation of the Propaganda in Rome. Pope Leo XIII's secretary asked the group for more information regarding its plea, which led the group to feel hopeful that its cause would be fairly and finally adjudicated.

It was at this point that Reverend Briscoe at St. Anne's went on a two-month vacation to the Holy Land, which the Notre Dame parishioners believed was really a trip to Rome on behalf of the bishop. Narcisse Rodolphe Martineau, a liquor dealer, was then elected to go to Rome on behalf of Notre Dame parish and had two audiences with the pope, two hours in all, coming back with sealed papers for Archbishop Williams in Boston and the belief that the pope had ruled in the parish's favor.

Hendricken then sent an Irish priest to perform Mass at Notre Dame, and the crowd grew so upset that police had to be called. Again, the church was shuttered.

The bishop met with Archbishop Williams and then came to Fall River to personally open the church. An Irish priest, Peter Feron, assisted him. One thousand Notre Dame parishioners refused to attend the Mass. The next Sunday, the bishop again appeared and read a letter from the archbishop that invited a delegation from Notre Dame to meet in Boston for a conference over the matter. The Sacred Congregation of the Propaganda had instructed Williams to grant Notre Dame its French Canadian priest, but he wanted to know whether the congregation saw this a right or a favor and whether it would submit an apology for its past transgressions and behavior.

The delegation was firm in its reply that the appointment of a French Canadian leader was its right and that the members of Notre Dame owed no one, not even the Catholic Church, an apology. And so the parish waited for a ruling.

In late 1885, Hendricken published a letter by Archbishop Williams that said that the Bishop could appoint anyone he wanted, provided he also appointed a French Canadian assistant. Father Feron was appointed curate and Reverend J.M. Laflamme was named as his assistant. At first, the congregation was incensed at this decision, but it then came to realize that Laflamme had been a pastor for a dozen years in a Quebec church and was older than Feron. The congregants deduced that it was just a matter of the bishop saving face and that, in a short while, they would get what they had been demanding. And so it came to be that on March 21, 1886, Reverend J.M. Laflamme was promoted as curate for Notre Dame parish, ending "the Flint affair."

In 1888, Laflamme resigned his parish, and the Reverend J.A. Prevost succeeded him. Notre Dame was then the largest French Canadian parish in the diocese of Providence.

Reverend Prevost found the original parish buildings to be old and dilapidated and not worthy of this flock, so he embarked on an enormous mission to build a new and suitable house of worship. Most parishioners were millworkers. For a sixty-hour week of work, wrote Philip Silvia Jr., "the average wage was $8.76. Yet over 3,000 families gave $1 per week on a voluntary basis," raising $215,000.

Architect and parishioner Louis Destremps worked with Prevost to design an edifice like no other seen in Fall River. Instead of a large church with an interior filled with supporting pillars that obstruct the view of services, this building would employ an innovative system of trusses, buttresses, beams and metal rods with turnbuckles so that there would not be a single interior column. To ensure this perfect view, an inclined floor with a three-and-a-half-foot pitch from the doors to the altar was employed. Said fire chaplain Thomas Harrington, "A tennis ball could easily roll down the main aisle to the altar rail in front."[109]

When the plans were presented to Bishop Harkins, successor to Hendricken, who died in 1886 at the age of fifty-nine, he thought the grand plans would mean expensive construction costs and asked them to reduce the size of the building, which they did. The final plans called for a Romanesque Corinthian building 235 feet in length and 100 feet in width, with twin steeples of 310 feet in height. The cornerstone was laid on May 30, 1891. Work began soon after to raise the granite walls, with stonecutters earning $1.50 per day and masons $1.25 per day for a ten-hour workday.

In 1893, the original wooden church was destroyed in a fire of suspicious origin. The new lower church was pressed into service in December 1894, and the following year saw the completion of all outside work on Notre Dame de Lourdes. According to Silvia, "Economic crises caused the delay between Notre Dame's foundation and its dedication in 1906—the national depression of 1893, and the strike of 1904, that put 20,000 Fall River millhands out of work for six weeks"; however, work proceeded on a new rectory and school. The interior of the upper church was begun again, and with the help of Canon Bouillon of the Archdiocese of Montreal, final designs for the interior decorations were made.

An artist and a sculptor were commissioned to produce the decorations—Mr. J. Castagnoli was to create all of the plasterwork on the interior, including columns, cornices and bas-reliefs, and Ludovic Cremonini was tasked with creating art to decorate the walls and vaults.

Cremonini began work on his murals, painted on canvas in sections, inside the church but could not work with onlookers coming and going and asking questions. Frustrated by his lack of progress due to these distractions, Cremonini had his work moved to a hall in the south end of the city. There, Cremonini worked and reworked his interpretation of the Last Judgment "in a tenement district where millhands raised their families, supported their parish, and pooled meager resources after long hours spent listening to the hypnotic drone of bobbins and spindles."[110]

He labored for almost three years until at last he completed his work. Said Cremonini, "I am not painting creeds and dogmas, I am painting mankind and history and art."

Painted in eight sections, it was the largest and most ambitious of the twenty-four works he created for the church. Cremonini's *The Last Judgment* measured seventy-five feet long and fifty-five feet wide. It included *God and the Fall of Lucifer*, *The Trumpeters*, *The Universal Resurrection*, *Christ and the Church*, *The Hebrews*, *The Temple* and *Charon, Cerberus, Minos, and Hell.*

> *The sections were joined in a successive motif on the platform of the nave, with all lines of juncture painted out. In the sanctuary, transept and cupola of the church, other works by Cremonini commanded parishioners to contemplate the love and wrath of God. They included 15 large paintings, depicting the mysteries of the Rosary, and in the platform of the sanctuary, five large works including the birth of the Virgin, the marriage of Mary and Joseph, and other scenes. Three extensive panels completed the work in the sanctuary, among them a painting of the Holy Family. Medallions of the four Evangelists, painted by M. Marrisael, flanked Cremonini's mural.*[111]

In addition to these grand works, a great organ was built and installed in 1906 by the Brothers Casavant of St. Hyacinthe, Parti Québécois, Canada, for a cost of $15,0000. The "high romance organ" had over three thousand pipes and seventy voices. The pipes were made of wood and metal and ranged in size from seven inches to sixteen feet in length.

Bishop Harkins dedicated Notre Dame de Lourdes on November 29, 1906. The building and decoration had cost $300,000. Prevost had raised $215,000 from parishioners and mortgaged the remainder, paying off the debt by 1923. Prevost announced at the dedication that thirteen years previously, he had been raised to the rank of apostolic protonotary with the title of monsignor. He had kept this news a secret until the day of dedication.[112]

In the 1924 jubilee year, the fiftieth anniversary of the founding of the parish, a nine-ton set of twenty bells from the Havard Foundry of Ville-Dieu, France, was installed in the south steeple. The five largest bells of the carillon were hung from yokes and could be sounded by hammers. Besides playing songs and pealing for services, the bells also struck the Westminster Chimes every fifteen minutes, which drew attention to the eight ten-foot clocks installed in the steeples the same year. When the carillon was blessed, each bell was given a name. The biggest was "Marie," in honor of the Blessed Virgin Mary, patron saint of the parish. "Marie" carried the following inscription in French: "My companions and I have been made to sing the praises of God."[113]

In addition to the carillon, the wooden floor in the aisles was redone in terrazzo and a special wool rug was woven for the sanctuary. Thirteen stained-glass windows, each measuring seven feet wide by twenty-seven feet high were also installed, created by the Maumejean Brothers of Paris. Smaller stained-glass windows were placed "in the sanctuary, sacristy, above the doorways, and above the interior cornice."[114]

A year later, Monsignor Prevost died. He was succeeded by Reverend Louis-Damase Robert in 1926. He is remembered for three major changes to the building, including the installation of a new acoustically correct pulpit, the ceiling of which could be adjusted to the preacher's height; the construction of a near perfect half-size reproduction of the Lourdes Grotto in the lower church; and lowering the church steeples by 75 feet.[115] In 1938, before hurricanes were named, one severely damaged both of the 310-foot steeples on Notre Dame. Bishop Cassidy directed that they be repaired and reduced in size to 235 feet.

When now Monsignor Robert fell ill in 1945, he was replaced by Reverend Philias Jalbert, who died prematurely in 1947, leaving the parish in "poor physical condition." The new pastor, Reverend Alfred J.E. Bonneau, served twenty-three years and completed major renovations without incurring any debt. He changed the coal furnaces to oil-burning systems, and when two hurricanes struck in 1954 less than two weeks apart, he had the severe damage to all of the parish buildings repaired.[116]

THE NOTRE DAME DE LOURDES FIRE OF 1982

In anticipation of the 100th anniversary of the founding of Notre Dame parish, the interior of the church was repainted, Cremonini's canvases were restored, new flood lights were added to the top of the cornice pointing toward the ceiling and others were installed on the lower window sills shining outward, raising the total number to 1,232.

Said parishioner Paul Chounard, "Walking into this church was like walking into heaven."[117]

In 1904, Pope Saint Pius X canonically erected the Diocese of Fall River, taking its territories from the Diocese of Providence. It now comprises the counties of Bristol, Barnstable, Dukes and Nantucket and three towns in Plymouth County. The first bishop was William Stang, who died in 1907.[118] In 1951, Bishop James Cassidy was its head, and in 1958, he raised Father Bonneau to the rank of monsignor. Monsignor Bonneau retired in 1970.

In the 1970s, the last major changes to the church took place. The lower grotto church had a seating capacity of 1,000, which was reduced, through the installation of a folding wall, to 475 seats. The rear portion became a hall and sat 400, with the organ room converted into a kitchen.[119]

At the time of the fire in 1982, Notre Dame de Lourdes was undergoing a $1 million renovation. Over $100,000 had been raised in cash, and another $400,000 had been pledged. The renovation drive had begun in 1978, with some of the 4,500 members of the parish giving $5 per week for that entire time. "It was the third major fire to hit the parish since 1968. Arsonists were blamed for the fires that destroyed Monsignor Prevost High School in 1968 and its annex in 1971."[120]

Roland Masse had been church sexton at Notre Dame for two years. His duties included maintaining the grounds, decorating for various occasions and laying out the vestments for services. "I was usually the first one in and the last one out," he said in an interview in 2012.[121]

Masse recalls that around 2:00 p.m. on Tuesday, May 11, he was closing the church windows when carpenters from Universal Roofing Company of New Bedford, who had been working on the repair and replacement of copper gutters, rushed into the building with horrible news. "They said, 'You have to get up to the attic. There's a fire in the attic.' Well, we didn't know how extensive the fire in the attic was."

Masse said he climbed about 120 feet of stairs with the workers to personally inspect the damage but had to turn around because "the smoke was just too thick." The sexton exited through the front door of the church, which faced Notre Dame and St. Joseph Streets, and looked up. He stated that the flames were not visible as they consumed the back east end of the church ceiling. "Everything looked great," Masse said. "I thought it was a smoky fire, not a big, raging blaze."

According to fire department records, the first alarm was called in a 2:25 p.m., having been placed from an on-site office trailer for Eastern Construction Company, which was handling the masonry re-pointing of the granite church. This was followed by another alarm at 2:27 p.m., a third at 2:29 p.m. and a general alarm, calling all firefighting apparatus and personnel to the scene, at 2:48 p.m.[122]

District chief John Furze was in charge at the signal station at Pine Street in Fall River. It was his job to call in and coordinate apparatus responding from neighboring communities. In all, there were some 350 firefighters and thirty-four fire departments from as far away as Norton, Brockton, Newport, Providence and parts of the Cape. There were fifty fire engines and fifteen or sixteen ladder trucks in use that day, "either directly fighting the blaze or in manning fire stations whose equipment had been sent to the scene."[123] By 5:30 p.m., nine lighting units were dispatched to assist the firefighters battling the blaze, which continued well into dark and included brush fires behind the Chace Mill on Rodman Street ignited from flying embers.

"First to respond was the 'Flint Task Force,' Ladder 4 and Engines 7 and 9."[124] Once the fire trucks arrived, says Masse, "I walked around the building to the east end and removed the Blessed Sacrament from the Tabernacle. The fire department was fearful the fire would spread to the rectory, about 45 feet from the church, so we evacuated the rectory of records, which went back 100 years, and everything we could get out of there. By the time I was done, the church was all in flames. I felt very bad [that] I could do nothing to save it but, really, there was nothing to be done."

Sister Carol Ann of the Sisters of Jesus-Marie was inside the church when the fire began, preparing forty-one children for their first communion. When she saw three workmen running for fire extinguishers, she alerted the other nuns in the building and calmly evacuated the children out the back door. By then, she said, the dome already was "ablaze, a bright orange ball of fire."[125]

Fall River fire chief Louis Shea said that the fire started when the wooden fascia board behind the gutters ignited during the construction work. The flames then quickly spread across the roof of the church and onto the twin steeples, which soon toppled onto the roofs of buildings that were next door, setting them on fire.

Said Deputy Chief Furze, "Firefighters were on the defensive from the word go. They say it spread faster than you could run."[126]

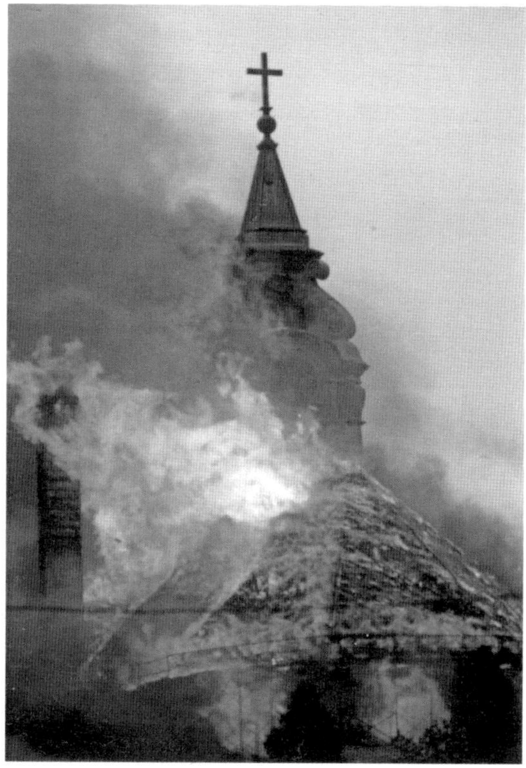

Right: Close-up view of the burning of Notre Dame de Lourdes church with the south tower in the background. It was determined that the fire started when workmen used a blowtorch to repair the copper gutters on the roof. *Michael Brimbau, photographer.*

Below: The church was not all that burned that day. Surrounding triple-decker homes and businesses also were destroyed in this conflagration, displacing three hundred residents. *Michael Brimbau, photographer.*

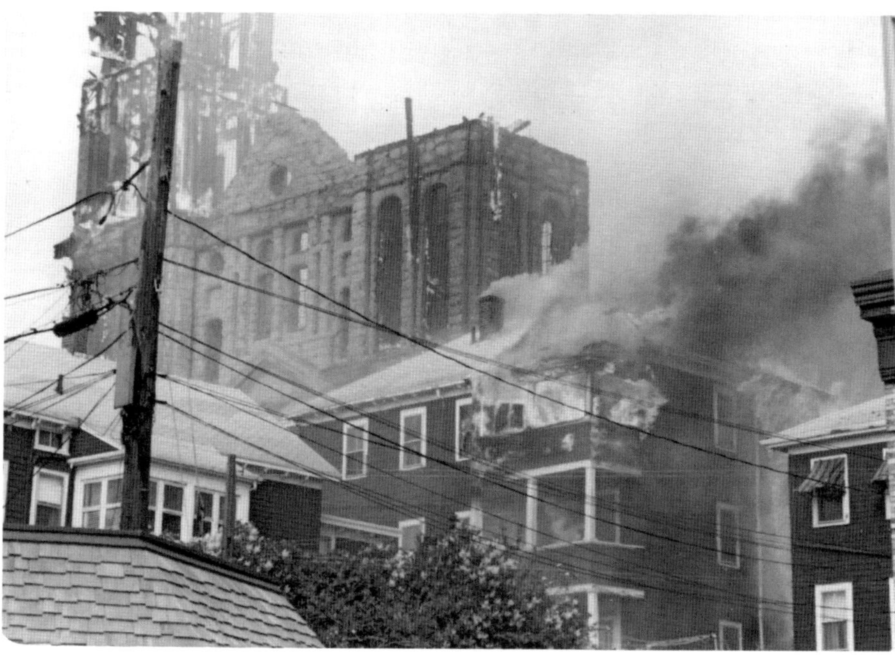

A strong wind was blowing from the north down, which spread the fire toward Pleasant Street. Additional factors at work in this fire included low water pressure, some faulty fire hydrants and an inability to communicate by radio with outside communities for assistance. Because Notre Dame is located in a tightly packed neighborhood, full of three-decker residences, and nearby the major business district of the Flint, there was the potential for a great loss of life. House after house was evacuated. "I was really worried," said Chief Shea. "Did we get everyone out? Was there anybody we didn't know about in any of those houses? Was there a blind person or a deaf person somewhere in one of those houses? It was frightening to even think that we might have missed someone." Shea says he felt lucky when all was said and done. "Think about it—it was 2:30 in the afternoon. Had it been 2:30 in the morning we would have lost 20 or 30 people."[127]

In broken English, an eighty-one-year-old Portuguese woman "said she couldn't leave her burning house because her husband was bedridden upstairs. The neighbors ran up, pulled the elderly man's trousers on, and carried him outside to safety."[128]

Miraculously, there was no loss of life, although fleeing residents lost virtually everything but the clothes on their backs.

The houses on Bedard Street were the first to go up in flames. Reports Thomas Harrington, "At the corner where Bedard Street intersected…the unthinkable was happening—the radiant heat from the southern exposure of the church was torching off an entire row of three-deckers…Oddly there was little smoke, just intense fire and heat." With great effort by the firefighters, one of the houses on Bedard was saved. When he realized that the other homes could not be saved, the chief had firefighters regroup on Pleasant Street.

Said Chief Shea, "You don't expect to fight a fire of this magnitude. I recall that I had just looked at some pictures of the Borden Mill fire, which I thought was bad. The Notre Dame fire made the Borden Mill blaze look like a weenie roast."

There were several close calls. District fire chief Edmund Cote was tasked with attacking the fire from inside the church. His crew of firefighters attempted to control the blaze, but it was soon apparent that this plan was hopeless. He ordered them to retreat. Two fire trucks that Shea had stationed on Bedard Street were nearly destroyed. The heat became so intense that Shea says, "I asked the deputy chief how long it would take for his men to uncouple from the hydrants. He told me, 'three minutes, tops.' I told him, to hell with that, drive out. Get out of there!" When the fire trucks roared away,

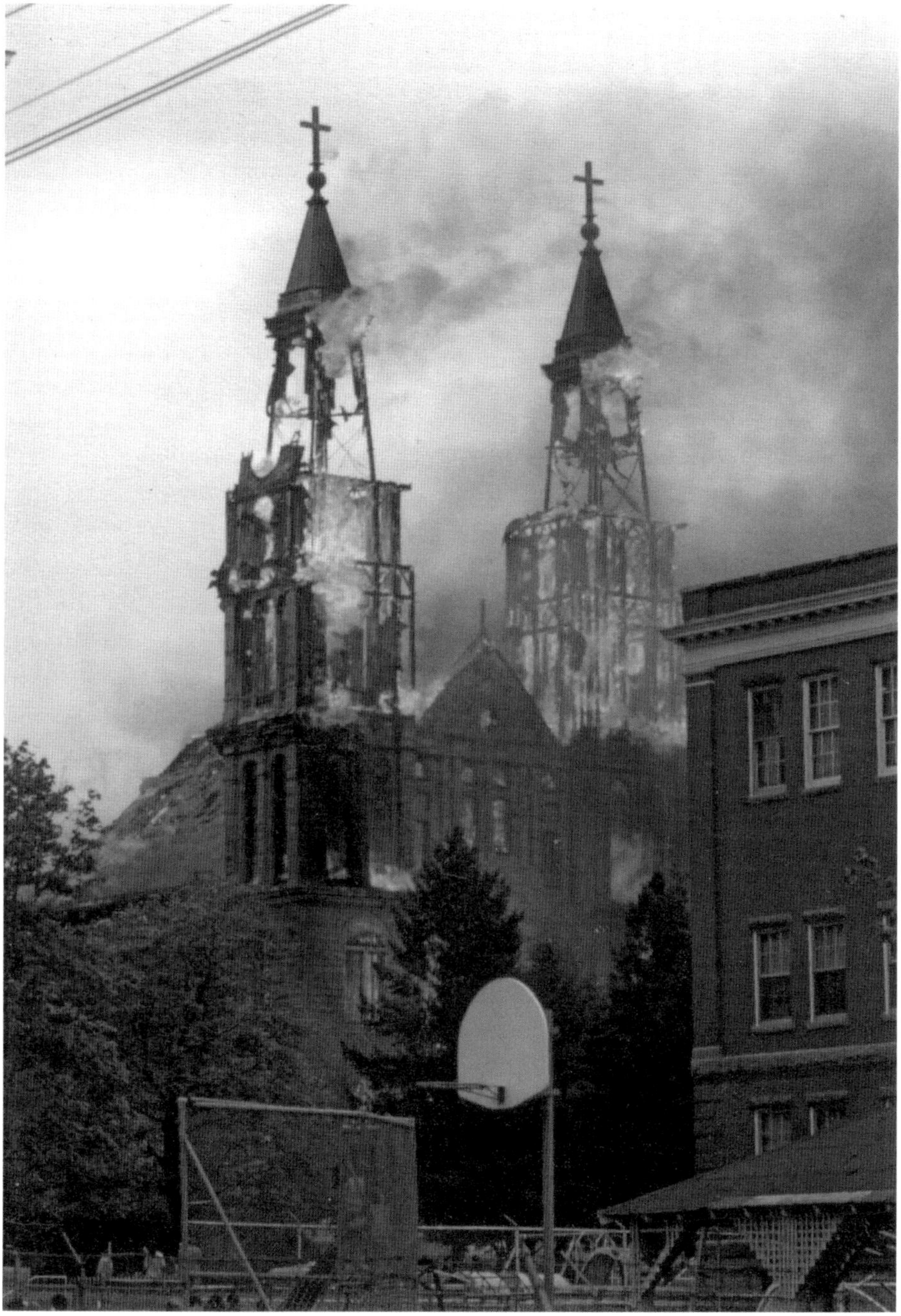

"It took one-half hour to level a church that took 16 years to build," said fire chief Louis Shea. *Michael Brimbau, photographer.*

they narrowly escaped a fireball heading in their direction. "The fire trucks left behind a trail of broken hoses, spewing water geysers several feet in the air as the lines snapped loose from the hydrants—but the firefighters were safe and, except for a paint blister or two, the engines were undamaged."

Reported Harrington, "There was at least one aerial ladder in place, and a heavy water stream was being directed by a firefighter belted in at the tip of the eighty-five-foot 'skick.' Handlines—big 2⅕ inch ones—and possibly a heavy-duty appliance with a couple of lines feeding it were pounding streams of water with lots of punch. None of this was making a dent."

One block south of the church, firefighters tried to stop the runaway blaze at Pleasant Street. "The next moments were probably the most dramatic I have ever experienced," said Harrington. "Everyone was tense. There were whispers that if we failed to halt the conflagration at Pleasant Street, the next line would be established all the way down at Alden Street by Espirito Santo Church and School." Despite the firefighter's best efforts, the fire eventually jumped the road and destroyed at least one commercial building on the other side. Said Shea, "At that moment I had all my pieces committed. I'm standing in the middle of Pleasant Street without any apparatus waiting for neighboring fire departments to come in. Houses were going up—Paradis Funeral Home was burning—a big, red brick building was burning on the other side. I was helpless."

Shea ordered a perimeter set up on Alden Street to be prepared in case a mill fire broke out. There was a call that a roof on the five-story Howard Arthur Mill was on fire. Apparatus from Warren and Bristol were sent there as soon as they arrived at Notre Dame. "They got there just in time."

Says Harrington, "I noticed several enormous stained glass windows along the north wall of the church lit up by the rolling flames within. Then, in an extraordinary turn of events, one of the windows burst and the square panes of stained glass fluttered down like a deck of cards dropped indiscriminately."

The heat from the fire was so intense and the winds blowing from the north so strong that this conflagration created its own wind system, what is known as a firestorm, a "vortex of wind and flame." As the heat from the fire drew in more and more of the surrounding cool air, an updraft was created. While firestorms do not ignite material itself, the intense heat sucks the moisture out of those materials and makes them more vulnerable to ignition by embers or firebrands, increasing the rate of fire spotting. Spewing embers, flaming debris and firebrands, the fire raced through Notre Dame, eventually felling the twin spires that had marked the city's skyline since

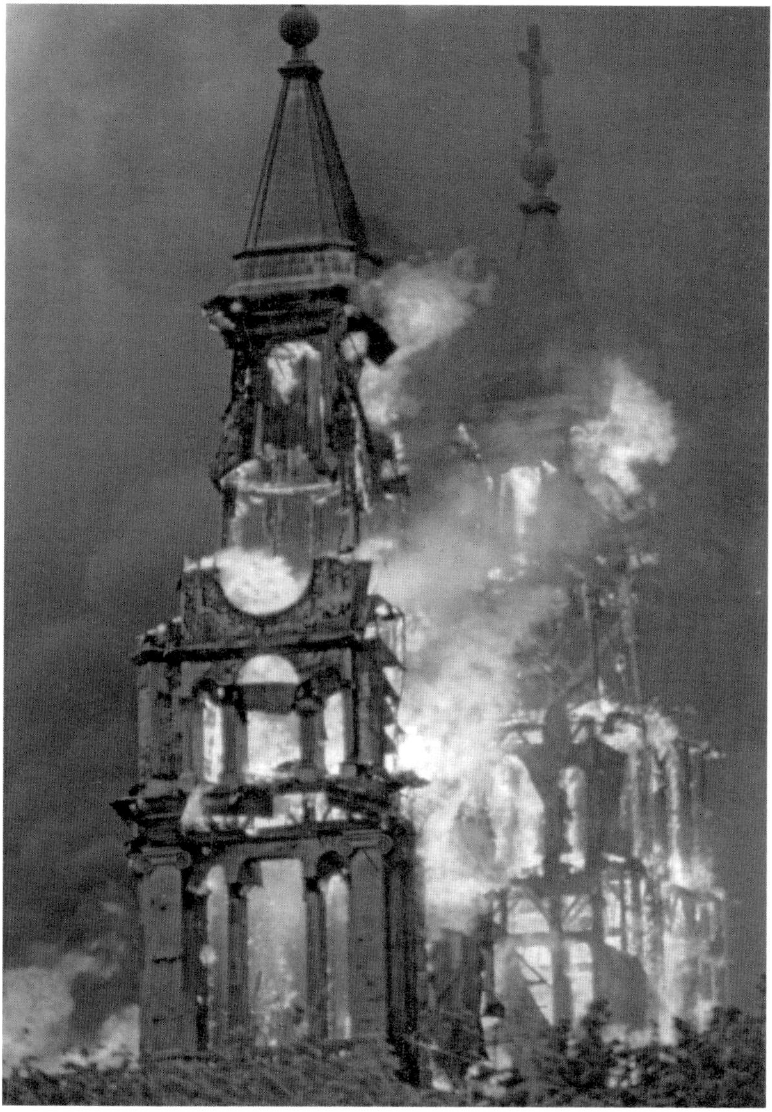

The twin 235-foot steeples of Notre Dame de Lourdes church just before they tumbled to the ground. *Michael Brimbau, photographer.*

1894. "By late afternoon, the church was a burning granite shell and the apartments nothing but smoldering rubble."[129]

> *The fire in the body of the church was finally abating now that everything combustible had been consumed. But it took one last wild and frightening*

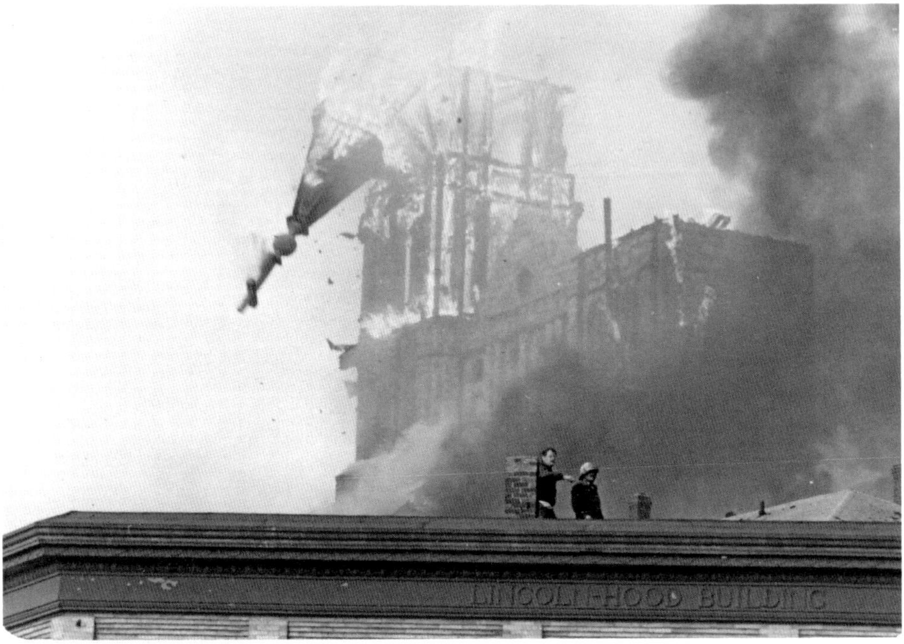

The toppling of the steeples of Notre Dame de Lourdes in 1982. *Michael Brimbau, photographer.*

> *turn when the elaborate twin towers succumbed to the flames, crashing and tumbling down, first the south, then the north. Spectators shrieked and cried. No one was remotely near the collapse zone and no lives were lost, but the psychological trauma and the symbolism of that final violent spasm horrified neighbors and parishioners.*[130]

The Notre Dame fire left three hundred people homeless and caused $50 million in damages, of which $20 million was estimated for the destruction of the church. Officials say 29 frame buildings, 109 apartment units and 8 small businesses were destroyed in the blaze. Another 3 residential and 12 commercial buildings were damaged. An undetermined number of buildings suffered roof and siding damage from flying embers and sparks.

Just one of the massive seventy-nine-foot-tall stained-glass windows was saved when the fire destroyed Notre Dame, only because it had been removed and sent out to be re-leaded.

The carillon, located more than one hundred feet up in the south steeple, fell when the massive wood beam and steel girder supports failed. Later, when wrecking crews were cleaning out the south steeple, they came across what was left of the bells. Since the bells had been made of bronze, an alloy that could not withstand the intense heat of this fire, they were melted into chunks. The steel clappers, on the other hand, and the yokes, or cast-iron frames, survived and were pulled out of the ruins of the fire.[131]

Fire chief Louis Shea believes that several factors were at play that, had they been different, could have saved Notre Dame and the surrounding structures. Several months previous to the fire, he had approved a $70,000 to $100,000 smoke-heat detector system in lieu of a sprinkler system. "If it had been in place," Shea says, "we would not have lost the church." The other issue was one that fire departments worldwide have to deal with and is, unfortunately, only too common a problem.

Notre Dame had been burning for at least an hour before anyone called the city's fire department. "At 2:05 p.m.," Shea said, "neighbors saw smoke coming from the roof—but nobody called the Fire Department. The primary problem in fighting that fire was the delay in reporting it. Fire had made considerable headway in the loft before we were notified. We were licked before we started. We had reports that people walking on Eastern Avenue saw smoke at 2:05 p.m. There is enough timber in that loft to fill two, maybe three lumberyards. It took one-half hour to level a church that took 16 years to build."

"It's the same old story," fire department training officer John Cashman said, gazing up at the stone façade. "The workmen spent about 20 minutes trying to put it out before calling the fire department."[132]

Reports were published in the local newspaper that the state's fire marshal's office would seek criminal complaints against Universal Roofing and Sheet Metal Company in connection with the conflagration. The contracting firm had been hired to repair and replace copper gutters on Notre Dame and had violated several state fire regulations. According to reports, the firm had used liquefied petroleum gas in tanks larger than the twenty-five-pound capacity allowed by law to fuel ovens and gas soldering devices. The state also alleged that the company failed to have suitable fire extinguishers immediately available and failed to take adequate measures to prevent ignition of adjacent combustibles. Conviction on charges of violating the state fire code carried a fine of $1,000.

Chief Shea felt that there was a very strong possibility that, in this case, there had been an hour-long pre-burn. The wood behind the gutter could have been exposed for a while to heat generated by soldering or related operations. "In such a case, a chemical decomposition of the wood begins, the ignition temperature is steadily lowered, the wood starts to glow and sets up the fire stage of 'flaming combustion,' the visible fire itself."

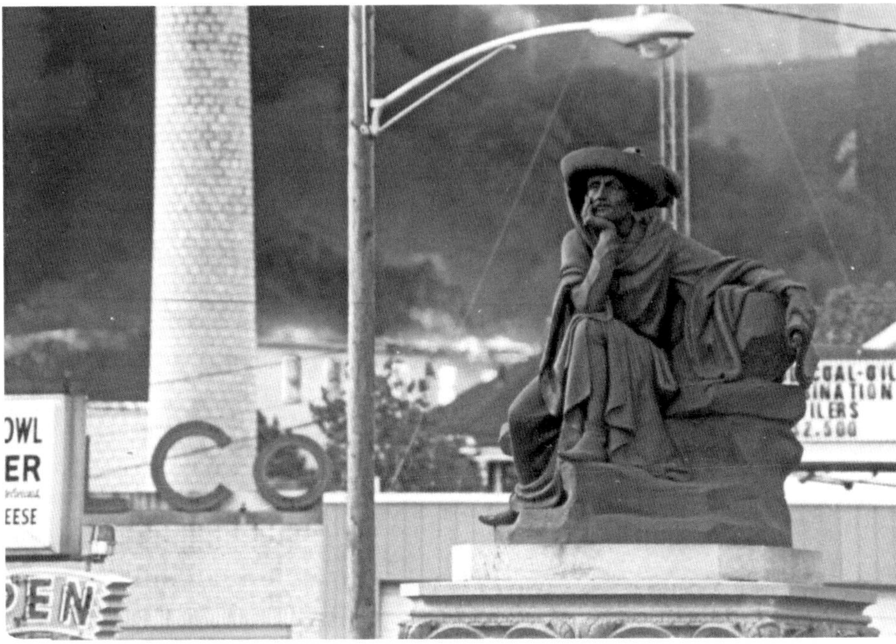

Photograph of the destruction of Notre Dame de Lourdes in 1982 as seen from the statue of *Prince Henry the Navigator* on Eastern Avenue. *Michael Brimbau, photographer.*

Notre Dame de Lourde's cornerstone was placed in 1891. While it sustained surface flaking because of the fire and was badly broken by wrecking equipment, it was discovered with its contents intact. In the capsule, not much more than a coffee can with its bottom rusted away, were found "two tiny statues, one of Our Lady and the other of St. Joseph, plus a medal. Also, there was a piece of parchment paper on which could still be made out the name of Louis Destremps as church architect."[133]

Postscript

THE KERR THREAD MILL FIRE OF 1987

No book titled *Historic Fires of Fall River* would pass muster if it did not include the Kerr Thread Mill Fire of 1987—a fire indelibly seared into people's minds not only because of its relatively recent occurrence but also because witnesses have such strong memories of it to this day. It was one hot fire and one huge event. Flames were visible from more than fifteen miles away in all directions.

Even though, thankfully, no one was killed or seriously injured, the human toll from this fire was significant. When the eleven light manufacturing businesses contained in the three large mill structures were destroyed, 950 people were left unemployed. When the fire raged out of control, 700 people were ordered evacuated from their homes as a precaution.

As many as five explosions, believed to have been caused by propane gas, sent flames as high as three hundred feet into the early morning sky. The intense heat of this blaze burned ten thousand feet of fire hose, caused granite curbstones to explode and liquefied macadam streets, as well as seared the exteriors and warped the vinyl sidings of some of the homes on the other side of the highway. "It looked like a tornado of fire," said ten-year-old Nancy Furtado. "It was hitting all the houses. I thought everything was going to go on fire."

Flames shot across the six-lane I-195, bombarding passing traffic with burning embers. Paint on cars bubbled, street signs melted, and windows cracked from the intense radiant heat.

Firefighters battled the blaze for almost five hours as the flames, fed by thirty-five-mile-per-hour winds and machine oil–soaked flooring, spawned a firestorm that threatened to consume additional buildings in its path. More than 185 firefighters from eighteen area communities assisted Fall River's 225 firefighters at the scene.

Homeowners with garden hoses watered down the houses on nearby McGowan and Jepson Streets, northwest of Interstate 195, to prevent the fire from crossing the highway.

Highways in the vicinity were closed for eight hours, including I-195 and Routes 6 and 24, and the Coughlin Elementary School on Pleasant Street was evacuated when it was learned that toxic chemicals were being stored in the basement of one of the buildings.

Within three hours, this incredibly fast-moving fire reduced the three large, six-story manufacturing buildings to empty shells, with only a few walls left standing and interiors in a fiery heap of debris. Damages were estimated to be $50 million.

And it was an expensive fire for the city of Fall River. Fire department personnel costs were $90,000 with replacement of equipment costing another $50,000, including $35,000 for the replacement of hoses. The Department of Public Works costs were $300,000, which included the replacement of over 1,200 feet of Martine Street, including a six-inch-deep street bed for the road, sidewalk and curbing. To level the buildings, $578,000 was spent removing material and hazardous waste and filling in at the site. Police department overtime and dedicated personnel costs amounted to $54,000. Total water department costs were $25,000, which included eight million gallons of water pumped by firemen and another eight million gallons pulled from the South Watuppa Pond to fight the fire and wet down the site after the fire.

This is how fast it happened.

The fire was officially reported at 7:11 a.m. Fire chief Louis Shea said that the fire department received six phone calls from neighbors on McGowan Street who reported thick, black smoke coming from the first and second floors at the west end of Building No. 3, which runs to the Brayton Avenue Extension. The box alarm came in almost simultaneously. The second alarm sounded at 7:24 a.m., quickly followed by the third. Shea ordered a general alarm at 7:24 a.m. and put out an assistance call for fire departments in neighboring towns.

At 7:34 a.m., 440-volt power lines near the burning mill exploded, sending a shower of sparks across the highway. Power was not restored to the 1,700 Westport residents who lost electricity until 7:00 p.m.

At 8:00 a.m., the second and third floors of Building No. 3 collapsed.

At 8:05 a.m., the west wall of Building No. 3 disintegrated, with other walls following suit shortly afterward.

At 8:07 a.m., the roof of Building No. 3 caved in.

The destruction of the other buildings soon followed.

By 11:30 a.m., the fire was under control.

At 4:40 p.m., police reopened the highway and allowed residents to return to their homes.

Fire officials said the blaze apparently started accidentally in the flue of the boiler in Building No. 3 and rolled through the mill like a large fireball of heated gases. Sprinkler systems worked properly but were quickly overwhelmed by the fire. After a three-and-a-half-month probe, with 110 witnesses and workers interviewed, arson was ruled out, and it was determined the boiler probably was never installed properly twenty years previously. The workers reported that there had always been boiler problems and that it had been extremely hot in the building before the fire, close to 90 degrees. Normally, it would take a long while to start up the boiler after the weekend, but on this Monday, it was already hot by 7:00 a.m.

The risk of fire in Fall River continues to this day, and its citizens are well aware of the disastrous possibilities facing the city. Modern firefighting techniques, equipment and safety gear are the only things standing between the residents and another major conflagration. Former fire chief Louis Shea's admonition is still, sadly, as germane today as the day he spoke it: "Fall River is a city built to burn."

APPENDIX

Population Figures for Fall River, Massachusetts

Year	Population
1810	1,296
1820	1,594
1830	4,153
1840	6,738
1850	11,524
1860	14,026
1865	17,491
1870	26,766
1875	45,340
1880	49,961
1885	56,870
1890	74,398
1895	89,203
1900	104,863
1910	119,295
1920	120,485

Year	Population
1930	115,274
1940	115,428
1950	111,963
1960	99,942
1970	96,898
1980	92,574
1990	92,703
2000	91,938
2010	88,857

The 1810–1930 population numbers are from Philip T. Silvia Jr., "The Spindle City: Labor, Politics, and Religion in Fall River, Massachusetts, 1870–1905." The 1940–2010 numbers are from United States Bureau of the Census.

"The Extent of the Calamity" of the Great Fire of 1843

Number of families residing within the burnt district at the time of the fire	225
Persons belonging to those families	1,324
Persons in addition, employed or doing business in the burnt district, but living out, about	600
Number of buildings burned, not including the smaller ones	196
Of which there were used as dwelling-houses and occupied by one or more families each	95
Hotels	2
Churches	3
Cotton Factory	1
Carriage Factories	2
Banks	2
Cabinet Warehouses	3
Marble Factory	1
Tannery	1

Livery Stables	4
Dry Goods establishments destroyed	17
Grocery and Provision establishments, including 3 or 4 Crockery Stores connected	24
Boot and Shoe Stores	6
Hat and Cap	3
Book and Periodical	3
Hard Ware	3
Milliners' Shops	11
Mantuamakers' Shops	5
Apothecaries'	6
Jewellers'	3
Harness Makers	3
Stove and Tin Ware	3
Brass Founderies	2
Blacksmiths' Shops	3
Machine Shops	2
Carpenters' Shops	8
Reed Makers' Shops	1
Shoe Makers' Shops	7
Plane Makers' Shops	1
Roll Coverers' Shops	1
Turners' Shops	1
Painters' Shops	8
Batchers' Shops	4
Soap Boilers' Shops	1
Segar Factory	1
Restaurators	7
Bake Houses	2
School House	1
School Rooms beside	3
Athenaeum	1
Customhouse	1

Post Office	1
Auction Rooms	1
Counting Rooms	7
Dentists'	2
Stage Office	1
Printing Offices	3
Lawyers' Offices	5
Physicians' Offices	5
Barbers' Shops	3

Note: All the merchandize shops in the place were destroyed, except some six or eight (principally groceries), situated in the extreme parts of the village.

Printed verbatim from An Authentic Account of the Value of Property Destroyed by the Great Fire in Fall River, July 2, 1843. *Boston: Dutton and Wentworth, Printers, 1844.*

The Granite Mill Fire of 1874, Dead and Wounded

Dead
Honora Coffee
Thomas Cavency
Catharine Connell
Maggie Dillon
Albert Fernly (eleven)
Gertrude Gray
Ellen J. Hunter
Mary Haley
Arabella Keith
Mary Lesondre
Bridget Murphy
Kate Murphy (sixteen)
Margaret Murphy (twenty)
Frederick Portrais
Annie J. Smith
James Newton (ten)
James Smith
Victoria Warner (twelve)
James Turner
Margaret Harrington
Mary Healy
(two missing)

Wounded
Mary Burns
Hannah Stafford
Mary Brodeur
Margaret Sullivan
Julia Coffee
Catharine Sullivan
Annie Dailey

Catharine Smith
Annie Healy
Annie Twoney
Joanna Healy
Margaret Twoney
Catharine Harrington
William Vinnecomb
Ellen Hanley
Delia Warner
Joanna Hanley
Mary Wriggley
Margaret Lanergan
Bethiah Wordell
James Mason
Ellen Jones
Julia Mahoney
Delia Portrais
Isabella Morehead
John Brodeur
Nancy Mullens
Thomas Gibson
Annie O'Brien
Edson Keith
Margaret Sullivan
Joseph Ramsbottom
Annie Sullivan
(two firemen)

From History of the Granite Mill Fire. *Fall River: Slade & Franklin, 1874, 17.*

Businesses Damaged and Destroyed in the 1916 Steiger Store Fire

Address	Business/Office	Extent of damage
South Main Street, West Side		
205	Baptist Temple	Water damage
209	R.E. McGuire Store	Smoke and water damage
221	Cobb, Bates & Yerxa	Total loss
229	Borphy & Hanley, clothiers	Total loss
231	J.F. Mull & Company, jewelers	Total loss
	Ovide Marquis, optician	Total loss
	Drs. Sears and Chisholm, dentists	Total loss
	Dr. J. Masterton, dentist	Total loss
	Abraham Shapiro, jeweler	Total loss
	Knights of Columbus, Fall River council, No. 86, hall and quarters	Total loss
233	S.S. Kresge's Five and Ten Cent Store	Total loss
251	Steiger Store	Total loss

APPENDIX

Address	Business/Office	Extent of damage
263	David Belford, tailor	Total loss
265	Frank M. Burgess, tobacco	Total loss
269	D.D. Sullivan & Sons, undertakers	Total loss
271	Joseph Winarsky, painter	Total loss
273	Morris Nariness, jeweler	Total loss
275	The Raiment, ladies clothiers	Total loss
281	New York Cloak & Suit Company, Cherry & Webb, proprietors	Total loss
289	Fall River Wholesale Dry Goods Company	Total loss
	Dr. P.W. Boylan, dentist	
	George F. Sullivan, tailor	
	Columbia Candy Company	
	National Truck Company	
	Rubinstein Brothers, shirt makers	
	Michael J. Daley, massage	
297	Madame Corneau, milliner	Total loss
303	The Fashion	Smoke and water damage
	Dr. J.J. Walsh, dentist	
	Dr. J.A. Burns, dentist	
	Dr. J.J. Kerrigan, specialist	
	South Main Street, East Side	
216	James Duckett, tobacco and steamship tickets	Smoke and water damage
218	Dr. A. St. George, dentist	Smoke and water damage
220	New York Shoe Shining Parlor	Smoke and water damage
224	John T. Sullivan, sample shoe	Total loss
228	Dr. Nathan H. Turk, dentist	Total loss
230	Charles F. Lavagnino, fruit	Total loss
234	Anna E. Holland & Company, millinery	Total loss
238	Benjamin Shogam, jewelry and musical instruments	Total loss
248	James C. Brady, druggist	Total loss

Address	Business/Office	Extent of damage
258	New York Clothing Company	Total loss
262	The Lenox Hotel	Total loss
264	New York Fur Manufacturing Company	Total loss
268	Rosa Soforenko, tailor	Total loss
270	Pawtucket Remnant Company	Total loss
272	Louis Tillis, tailor	Total loss
276	Michael H. Reed, barber	Total loss
280	Robert J. Murphy, lunch room	Total loss
282	David Wolk, stationery and dry goods	Total loss
286	Shoe Shine Parlor, Veras and Poppas	Total loss
288	Fruit Store, Veras and Klondas	Total loss
296	Louis Madowsky, notions	Smoke and water damage
300	Connelly and Murphy, wholesale liquors	Smoke and water damage, plate glass window smashed
	SPRING STREET	
385	P.P. Morris, paint shop	Total loss
382	Star Cleaning Company	Total loss
384	Kilroy Brothers, furniture	Total loss

From Fall River Evening Herald*, "The Losers," February 16, 1916.*

BUSINESSES DESTROYED IN THE GREAT FIRE OF 1928

ADDRESS	ROOM	BUSINESS/OFFICE
		SOUTH MAIN STREET, WEST SIDE
1		George A. Lake, jeweler
		G. Foxhall, optometrist
3		James Gianakos Company, bootblacks
5		United Cigar Store
7		Vacant

APPENDIX

Address	Room	Business/Office
13		Joseph A. Bowen & Company, coal
15		Thomas G. Barlous, restaurant
19		Conos Brothers, confectionery
21		Samuel H. Allen
23		Granite Block
	Room 1	Edwin J. Mills, deputy sheriff
	Rooms 4–6	Arthur J. Doucet, real estate and insurance
		Oliva St. Denis, real estate and insurance
	Room 5	Charles P. Ryan, lawyer
	Room 9	James E. Cunneen & Company, insurance
	Room 11	D.T.L. Castellano, tailor
	Room 12	Nicholas Hatheway, lawyer
	Room 14	Fall River Real Estate Association
	Room 15	Severin M. Lamarre, lawyer
		Albert A. Belanger, publisher
		Eugene S. Sullivan, deputy sheriff
	Room 17	Almel L. Audet, real estate
	Room 19	John W. Cain, insurance
	Room 20	Frank A. Pease, lawyer
	Room 21	Thomas F. Higgins, lawyer
	Room 22	Frank Amber, barber
	Room 23	William A. Torphy, lawyer
	Room 25	Herman J. Legare, auctioneer
		Hubert Legare, insurance
		Mariano Moniz, real estate
	Room 26	Joseph Roderick, lawyer
	Room 27	Herbert B. Winchester, watchmaker
	Room 29	Joseph Menard, insurance
		Raymond T. Menard, insurance
		Amable Chouinard, real estate
	Rooms 31–35	William J. and J.H. Hickey, real estate and insurance
		U.S. Casualty Company

APPENDIX

Address	Room	Business/Office
	Room 101	Margaret J. Howarth, lawyer
	Room 102	Anthony Alves, real estate
		Edwin Belcher, insurance
		W. Harry Monks, real estate
	Room 103	Charles S. Graham, auctioneer
		Graham Loan Company
		Lake Mailing Company, office supplies
	Room 104	The Exchange, real estate
	Room 105	Nathan Sternsher, lawyer
	Room 109	Fernald L. Hanson, lawyer
	Room 110	Thomas D. Sullivan, lawyer
	Room 112	Romauld J. Crispo, lawyer
	Room 115	William F. Fuller, lawyer
		L. Elmer Wood, lawyer
		Brayton Morton, lawyer
		George L. Sisson, lawyer
	Room 116	Harold E. Clarkin, lawyer
	Room 121	Walker Typewriter Company
	Room 123	Benjamin Cook Jr., lawyer
	Room 124	Cook Taylor Company, Inc., mill supplies
		Napoleon Jasmin, real estate
		Joseph H. Levesque, real estate
	Room 125	William Lambert, real estate
	Room 127	Frank L. Piper, lawyer
		Fred O. Dwelly
	Room 130	John T. Farrell, lawyer
		Frank A. Sullivan, real estate
	Room 201	Harry Greenwood
	Room 207	James F. Mooney, contractor
	Room 209	Cash Brothers Inc., roofers
		Luxor Cab Company, taxi service
	Room 211	Frank H. Sullivan, real estate and insurance

APPENDIX

Address	Room	Business/Office
	Room 213	Bernard A. Doherty Jr., lawyer
	Room 217	Lorenzo E. Savard, real estate and insurance
	Room 219	Mrs. E.V. Cloutier, chiropodist
	Room 221	Cobb, Bates and Yerxa Company, grocers
	Room 223	Baldwin's Credit Exchange
	Room 225	Alice Botway, dressmaker
	Rooms 226–31	C. Cockroft and Company, contractors
		Bergeron and Bergeron, lawyers
		Frank X. Perron, real estate
		Thomas F. Powers
27		Thomas A. Borden, stationer
29		P.S. Brown, druggists
33		Michael A. Clorite, fruit
35		Waiting room, News Company
		George Collias, bootblack
		Eastern Massachusetts Street Railway Company Information Bureau
41		Central News and Delicatessen
47		James C. Brady, druggist
SOUTH MAIN STREET, EAST SIDE		
6		Citizens' Savings Bank
8		Citizens' Savings Bank Building
	Rooms 21–24	Swift, Grime, Buffington and Crossley, lawyers
	Room 25	William L. Harrison, real estate and insurance
	Room 30	D.W. Riley and Son, real estate
	Room 31	William F. Thomas Jr.
	Room 33	David Shapiro, architect
	Room 35	Penn and Pallette, commercial artists
	Room 36	Service Printing Company
20		Union Savings Bank
NORTH MAIN STREET, WEST SIDE		
7		Stavros and Company, fruit

Address	Room	Business/Office
9		Eagle Restaurant
11		Burke and Davis, men's furnishings
17		Majestic Restaurant
23		Arthur H. Rodgers, men's furnishings
25–33		New Wilbur Hotel
33		The Norfolk Hosiery Company
NORTH MAIN STREET, EAST SIDE		
2–28		Durfee Block
2		Mohican Drug Company, Inc.
6		The Mohican Hotel
		William Durfee
		Mohican Theatre Company, Inc.
		Pacific Oil Company
		Durfee and Canning Company
		Matthew J. Mansour, cigars
12		Dennis N. Sullivan, men's furnishings
16		City News Company, news dealer
20		C. Warburton and Sons, florists
24		Crystal Restaurant
28		Rialto Theatre
32		E.J. Sokoll Company, confectioners and caterers
34		Merchants Block
	Room 1	Lucio Parise, tailor
	Room 2	Patrick J. Keating, barber
	Room 5	John W. Beattie, real estate
		Edwin N. Blinn, lawyer
		Edward P. Fitzgerald, typewriters
		Alexander Lockhart, real estate
		Irving L. Mayo, painter
	Room 6	Anthony Brothers, music publishers
	Room 7	Jacob Kalevich, tailor
	Room 8	Dana C. Everett, cotton broker

APPENDIX

Address	Room	Business/Office
	Room 11	McCartys' Band and Orchestra
		John J. McCarty, piano tuner
	Room 21	Fall River Publicity Bureau, publishers
		Victor S. D'Unger, publisher
36		C.E. Gifford and Company, jewelers
38		Astor Lunch
PURCHASE STREET, WEST SIDE		
7		Direct Sales and Finance Company, cotton
11		Troy Co-operative Bank
17		Grandfield and Mycock, cloth brokers
		George E. Brown, cotton merchant
21		E.H. Brow and Company, cloth brokers
		C.D. Davol and Company, cloth brokers
25	Room 1	A.G. and W.J. Howland, real estate and insurance
	Room 2	Capital Service Company, cotton
	Room 3	Hathaway & Company, cotton
	Room 4	Bryant Chapin, artist
		R.M. Deane and Company, cotton brokers
	Top floor	Gagnon Printing Company
		LePetit Courier
27		Ann's Kitchen, restaurant
PURCHASE STREET, EAST SIDE		
10		Buffington Building
	First floor	James Buffington, cotton cloth
		Aldrich & Hawkins, cotton
		Lesser-Baldwin Company, cotton
	Room 21	Tonkonogey & Adler, real estate and insurance
		Fall River Granite & Quarry Company
	Room 22	Myer Markell, real estate
		Maurice Gomberg, insurance
		Lester Bakst, lawyer
	Room 24	Arthur S. Phillips, lawyer

Address	Room	Business/Office
		Edward J. Lajoie, lawyer
	Room 26	James N. Buffington & Company
	Room 27	Harry L. French, accountant
	Room 31	Edward T. Robertson & Son, cotton controllers
	Room 32	Daniel J. Harrington, lawyer
	Room 34	William O. Manion, insurance
		Universal Thrift Stamp Company
	Room 36	Coughlin & Harrington, lawyers
		Frank J. Andrade, constable
	Room 41	Massachusetts Income Tax Office
	Room 42	Income Tax Assessors
	Room 44	Henry M. Fenner, insurance
	Room 45	Fall River Historical Society
	Room 46	Aetna Life Insurance Company, claim department
	Room 47	Swift and Apperson, accountants
	Rooms 51–53	Ralph E. Loper and Company, industrial engineers and accountants
	Room 54	Taxpayers Association of Fall River, Inc.
	Room 56	Fall River Board of Underwriters
		New England Insurance Exchange
BEDFORD STREET, SOUTH SIDE		
27	Second floor	Textile Trucking Company
		Fred J. McLane, real estate
		U.S. Naval Reserves
	Third floor	Clifford M. Gardner and Company, cotton brokers
	Fourth floor	Burns Club
29		Massasoit-Pocasset Bank Building
		Lincoln and Hood, lawyers
		Harold Hood, lawyer
		Preston Hood, lawyer
		A.H. Hood and Company, general insurance
	Room 25	Ernest A. Lincoln, civil engineer

APPENDIX

Address	Room	Business/Office
	Room 27	Fall River Co-operative Bank
	Rooms 33–35	Baker, Seagrave and Terry, lawyers
	Room 37	Dr. Harold L. Stover, dentist
	Room 38	Orient Laplante, lawyer
	Room 41	Booth and Chase Company, Inc., contractors
	Room 43	Patrick J. Hurley, office
	Room 44	Burroughs Adding Machine Company
	Room 47	George R. Lawton, accountant
	Room 53	John Hancock Mutual Life Insurance Company
	Room 58	Cotton Manufacturers Association
33		Massasoit-Pocasset National Bank
BEDFORD STREET, NORTH SIDE		
0		Checker Taxi Company
2		William C. Atwater and Company, Inc., coal
		Fall River Navigation Company, Inc.
12		Smith Office Equipment Company
14		Western Union Telegraph Company
16		Stephen M. Weld and Company, cotton dealers
		The Reilly Press, printers
20		Staples Coal Company of Fall River
22		Greene and Son, real estate and insurance
		Francis T. Borden, civil engineer
24		Birch Building
	First floor	Sanford and Kelley, stock brokers
		Brayton and Osborn, Inc., cotton
	Second floor	Roland G. Adam, tailor
		Franklin Gray and Company, cotton
	Third floor	Katherine L. Ney, teacher of dancing
28		The Munroe Press, printers
32		John F. Stafford and Company, real estate and insurance
58		George N. Durfee, cotton broker

APPENDIX

Address	Room	Business/Office
		Edgar G. Durfee, cotton broker
		Mrs. Anita Dionne, hairdresser
		Edmond A.R. Grenier, barber
		New York Life Insurance Company
		Frank A. Flynn, insurance
60		George N. Durfee and Sons, insurance
		People's Co-operative Bank
		William L. Cooper
64		G.M. Haffards and Company, brokers
		Clinton V.S. Remington, cotton cloth and yarns
		Robert K. Remington, cotton cloth
68		George H. Waring and Company, yarn department
		George H. Waring and Son, cotton brokers
70		Waring Building
	Second floor	Frederick J. Beckett, real estate
		Edward M. Corbett, civil engineer
		Fall River Blue Print Company
		Highland View Land Company
		New Boston Road Land Company
		William O. VanBlarcom
	Third floor	N.R. Earl and Son, bookbinders
74–76		E.A. Shaw Company, Inc., cotton dealers
78		Metacomet Bank Building
	Room 3	Stephen Apps and Company, cotton cloth
	Room 4	Dr. William J. Speers, dentist
		Dr. Cornelius Weeks, dentist
	Room 8	Frank N. Davis, mill supplies
	Room 9	King Philip Mill office
	Room 11	J.H. Franklin and Company, printers
	Third floor	Clan McAlpin Hall
86		Metacomet National Bank
102		Francis P. Sarmento, restaurant

APPENDIX

Address	Room	Business/Office
106		Bay State Rooming House
		Charles F. Curtice
108		William F. McCoy, cotton
110		Euclide J. Duquette, barber
116		Robert O. Wilcox and Company, paper
118		Sophia St. John, boarding house
GRANITE STREET, SOUTH SIDE		
21		Printers' Hall
		Fall River Poster Advertising Company
		J.W. Burrell and Company, cotton brokers
		Arthur J. Sullivan, sign painter
25		McDermott Lunch
39		Smith Oil Company, Inc., filling station
77		Norman G. Hicks, electrical contractor
		Standard Vulcanizing Company
GRANITE STREET, NORTH SIDE		
6		Etienne Gagne, barber
16		Flossie's Kitchen, restaurant
20		Vacant
22		Boston Advertiser
		Boston American
		Providence Journal
28		City Building Department Garage
38		Vacant
40		Vacant
46		Vacant
82		Abel H. Wordell, plumber
BANK STREET		
165		Congregation Beth-El
POCASSET STREET		
		Pocasset Mill
203		Herald News Building (partially damaged)

Address	Room	Business/Office
		Fall River Daily Herald Publishing Company
		Herald News
207		Vacant
211–13		Nathan Miller and Sons, Ltd., plumbers
		Old Herald News Building
219–21		L'Independent Publishing Company
		Jean B. Paradis and Company, real estate
ROCK STREET		
18		Premier Theatre

From Pictures of the Worst Fire in Fall River's History*, New Bedford, MA: George H. Reynolds, 1928, and* Fall River, Massachusetts, City Directory, 1928.

Bishops of the Roman Catholic Diocese of Fall River

William Stang	1904–07 (died)
Daniel Francis Feehan	1907–34 (died)
James Edwin Cassidy	1934–51 (died)
James Louis Connolly	1951–70; retired 1970, died 1986
Daniel Anthony Cronin	1970–92, appointed archbishop of Harford, Connecticut, retired in 2003
Sean Patrick O'Malley, OFM Cap	1992–2002, appointed archbishop of Boston in 2003, appointed cardinal-priest in 2006
George William Coleman	2003–14, retired
Edgar Moeira da Cuhna, SDV	2014–present (only Brazilian-born bishop in the United States)

NOTES

CHAPTER 1

1. See Appendix, Population Figures for Fall River, Massachusetts.
2. *Fall River and Its Manufactories*, 13.
3. Fenner, *History of Fall River, Massachusetts*, 16.
4. Belanger, *Guide to Fall River's Mills*, v.
5. Peck and Earl, *Fall River*, 7.
6. Lima, Castro and Champlin, *River and Its City*, 103.
7. Earl, *Centennial History*, 10.
8. Lima, Castro and Champlin, *River and Its City*, 112–13.
9. Fowler, *History of Fall River*, 72.
10. Bronson, *Address on the Anniversary of the Fire*, 3.
11. Fowler, *History of Fall River*, 72.
12. Details in this chapter regarding Fall River firefighting and apparatus history, especially those used during the fire of 1843, are found in Lynch, *History of the Fire Department*, 87–93.
13. AuroraRegionalFireMuseum.org.
14. Lynch, *History of the Fire Department*, 10.
15. Ibid, 88.
16. Bronson, *Address on the Anniversary of the Fire*, 4.
17. *Fall River All Sorts*, July 11, 1843.
18. Fowler, *History of Fall River*, 73.
19. Flanigan, "Fire of 1843," 5.
20. Martins and Binnette, *Parallel Lives*, 121.

21. *Fall River All Sorts*, July 11, 1843.
22. Fowler, *History of Fall River*, 77.
23. *Authentic Account*, 31.
24. See Appendix, "The Extent of the Calamity" of the Great Fire of 1843.
25. Bronson, *Address on the Anniversary of the Fire*, 7.
26. Fenner, *History of Fall River, Massachusetts*, 25.

Chapter 2

27. Belanger, *Guide to Fall River's Mills*, 25.
28. Lynch, *History of the Fire Department*, 118.
29. Belanger, *Guide to Fall River's Mills*, 25.
30. Lawlor, *Life and Struggles of an Irish Boy*, 28.
31. Higgins, *Comprehensive Dictionary of Textile Terms.* Carding is a process—performed by a carding machine—in the preparation of cotton and wool fibers that separates and parallels the fibers, removes impurities and delivers the fibers in sliver form. Sliver is the loose, rope-like untwisted fibers produced by the carding machine; roving is the product of the roving frame machine, the size of which is determined by the count of the yarn to be spun. A roving frame machine is used to reduce the cotton sliver to a size in ratio to the count of yarn required. This process precedes spinning.
32. Lima, Castro and Champlin, *River and Its City*, 119. Spinning is the final drawing and twisting of the cotton strand into a thread and the winding of the thread onto a practical device, such as a cone or spindle. Spinning is done either on the ring frame or the mule frame; Higgins, *Comprehensive Dictionary of Textile Terms.* A spinning mule is a machine that first spins the yarn and then winds it on a spindle into a small package termed a cop. It spins the yarn of any counts but is generally used to obtain very fine counts; since mule-spun yarn fibers do not lie straight in the yarn, as they do in ring-spun yarn, they form a bulkier yarn and thus a fabric that is fuller.
33. Lima, Castro and Champlin, *River and Its City*, 119. The drawing process occurs immediately following the carding and takes the loose, thick cotton stock and reduces the diameter of the strand by stretching it out through the use of rollers. The individual fibers are not stretched at all; the diminishing of the strand diameter results from the attenuation as it progresses through the rollers.

Higgins, *Comprehensive Dictionary of Textile Terms.* Spooling is changing into a larger unit or spool from the spinning bobbin. A slasher is a machine used for sizing warp yarn. The yarn is taken in an even sheet from the section beams; passed through the size liquid and through squeeze rolls; then wrapped around hot cylinders and fans, to dry the sheet; and finally pressed solidly onto the weaver's beam.

Lima, Castro and Champlin, *River and Its City*, 119. In warping, racks of bobbins are set up to hold the thread while it is rolled onto the warp bar of a loom. Because the thread is fine, often three of these would be combined to get the desired thread count. This operation is in preparation for slashing or dyeing.

34. Dunbar, "Disastrous City Mill Fire."
35. Lima, Castro and Champlin, *River and Its City*, 120.
36. Dunbar, "Disastrous City Mill Fire."
37. Lawlor, *Life and Struggles of an Irish Boy*, 24.
38. Greenlees, *Female Labour Power*, 107.
39. Lawlor, *Life and Struggles of an Irish Boy*, 29.
40. See Appendix, Population Figures for Fall River, Massachusetts.
41. Blewett, *Constant Turmoil*, 408.
42. *History of the Granite Mill Fire*, 29. While many sources, including Lynch, Blewett, Mitchell and Dunbar, place the origin of the fire on the third floor, the most contemporaneous accounting, and the official one, reports that the fire started on the fourth floor in the spinning room. The accuracy of this is further supported in that the mule spinning was performed on the fourth and fifth floors of Granite Mill No. 1, according to the same source as well as the account by Lawlor, a child at work in Granite Mill No. 2 at the time of the fire.
43. Unless indicated, all quotes regarding the fire are from *History of the Granite Mill Fire*.
44. Lawlor, *Life and Struggles of an Irish Boy*, 31.
45. McCreary's testimony is given in *History of the Granite Mill Fire*, 18–20.
46. Lynch, *History of the Fire Department*, 116.
47. *History of the Granite Mill Fire*, 29.
48. Dunbar, "Disastrous City Mill Fire."
49. *Fall River Daily Herald*, "Granite Mill Fire."
50. Lawlor, *Life and Struggles of an Irish Boy*, 31.
51. *History of the Granite Mill Fire*, 5–6.
52. See Appendix, The Granite Mill Fire of 1874, Dead and Wounded.
53. "Verdict of the Jury," *History of the Granite Mill Fire*, 28–32.
54. Dunbar, "Disastrous City Mill Fire."
55. *Fall River Herald News*, "Granite Mill Fire Resulted in New Laws, Regulations."

Chapter 3

56. These population figures are suspect and do not jibe with the census counts for 1920, which shows the population at 120,485, the highest number the city would attain (see Appendix, Population Figures for Fall River, Massachusetts).
57. The *Fall River Daily Globe*, the *Fall River Daily Herald* and the *Fall River News*. The French-language newspaper was *L'Independent*, and there were also two Portuguese weeklies at the time: *As Novidades* and *Michaelense O*.
58. *Fall River Globe*, "Costly Fire at American Linen Mills."
59. Goldberg's statements throughout this chapter come from the *Fall River Daily Globe*, "Ellis Goldberg Sent in First Alarm."
60. Phillips, "How Arson Investigation Has Changed." The modern science of fire investigation did not begin in America until the 1970s with the creation of the National Institute of Standards and Technology.
61. O'Rourke's statements come from three sources: *The Standard*, "Lessons of the Fall River Fire"; *Fall River Evening Herald*, "South Main Street Is Swept by Blaze; 23 Buildings Lost"; and *Fall River Evening Herald*, "Steiger Watchman Makes Statement."
62. *Fall River Evening Herald*, "South Main Street Is Swept by Blaze."
63. *Fall River Daily Globe*, "Loss of a Million and a Half by a Midnight Conflagration."
64. Ibid., "Fine Work by the Firemen of Other Cities."
65. *Fall River Evening Herald*, "South Main Street Is Swept by Blaze."
66. *The Standard*, "Principal New England Losses of the Week," 192.
67. *Boston Globe*, "Fall River Swept by $1,500,000 Fire."
68. This story, as well as the ones about Rogers and the blind man escaping from the Lenox Hotel, come from *Fall River Evening Herald*, "South Main Street Is Swept by Blaze."
69. *Fall River Evening Herald*, "Lenox Hotel Guests Have Narrow Escape."
70. Until indicated, the next series of quotes are all from *Fall River Daily Globe*, "Loss of a Million and a Half by a Midnight Conflagration."
71. *Boston Globe*, "Fall River Swept by $1,500,000 Fire."
72. Until indicated, the next series of quotes are all from *Fall River Daily Globe*, "Loss of a Million and a Half by a Midnight Conflagration."
73. *Fall River Evening Herald*, "Lenox Hotel Guests have Narrow Escape."
74. Ibid., "Thoughtful Merchants."
75. *Fall River Daily Globe*, "Cathedral Luckily Uninjured."
76. *Fall River Evening Herald*, "Cause of the Big Fire Not Yet Determined."

77. See Appendix, Businesses Damaged and Destroyed in the 1916 Steiger Store Fire.
78. *Fall River Evening Herald*, "Cause of the Big Fire Not Yet Determined."
79. Ibid., "Fully 20,000 People Attracted to Fire Zone."
80. *The Standard*, "Principal New England Losses of the Week," 192.
81. *Fall River Evening Herald*, "Steiger Going to Have Better Store."
82. All quotes related to the fire inquest are from *Fall River Evening Herald*, "Origin of Steiger Fire Is Mystery."

Chapter 4

83. See Appendix, Businesses Destroyed in the Great Fire of 1928.
84. Most often found at work sites, a salamander heater is a forced-air or convection heater, often fueled by kerosene. In this case, a fifty-gallon steel oil drum had one end removed and workers had been burning wood inside it as it rested directly on the oil-soaked flooring of the mill. These types of improvised heating devices should not be used inside buildings, where flying embers can cause fires.
85. According to *A River and Its City*, the seventh falls above tidewater were 21.67 feet and supplied 225.4 horsepower. The Quequechan River had a total of eight waterfalls, with a total minimum fall of 129.24 feet, providing 1,330.2 horsepower. The Pocasset Manufacturing Company had the highest fall and the most horsepower.
86. Information for this section on the history of the Pocasset Mills comes from Belanger, *Guide to Fall River's Mills*, 5; Lima, Castro and Champlin, *River and Its City*, 105, 108; Earl, *Centennial History*, 22–27; Phillips, *Fascicle II*, 119–22, 126; and Fenner, *History of Fall River, Massachusetts*, 58–60.
87. Lima, Castro and Champlin, *River and Its City*, 108. In block printing, a design in relief is carved out of a block of wood and a length of rolled cloth spread out across a table. Inked blocks are pressed against the cloth and struck by a wooden mallet to ensure a good impression. The cloth is then rolled forward over drying rollers and succeeding sections of cloth impressed with the blocks until the length of cloth is fully printed. Different colors require additional runs through the blocking process.
88. Ibid. Roller printing mechanized the printing operation and allowed the continuous printing of cloth from engraved copper rollers. The first machines of this type allowed textiles to be printed in up to six colors in a single operation. Roller printing now allowed large volumes of cloth to

be printed, and it has been the principal method of printing cloth during the past two hundred years. One of the challenges of early printing was the proper drying of the cloth. Large drying sheds were built adjacent to the Quequechan falls to dry the cloth in the air. When the weather was damp for several days, resulting in the slow drying of the cloth, the print works were forced to shut down. Later, machine drying of print cloth solved this problem.

89. Ibid., 122. Quequechan Mill: fall twenty-one feet, horsepower 218.4.
90. Ibid. Watuppa Mill: fall 15.38 feet, horsepower 160.
91. *Fall River Conflagration, February 2–3, 1928.*
92. *Fall River Herald News*, "Witnessed Great Fire."
93. *Fall River Conflagration, February 2–3, 1928.*
94. *Fall River Herald News*, "Her Jewelry Lost in Fire."
95. *Fall River Daily Globe*, "Madam's Dummies Go to Bank but Meet Harsh Cops"; ibid., "Spectators Think Dummies Are Bodies of Fire Victims"; *Fall River Herald News*, "Wax Figures Are Removed"; *Fall River Herald News*, "Odd Occurrence at Conflagration."
96. *Fall River Herald News*, "Witnessed Great Fire."
97. It exists in the same location to this day.
98. *Fall River Conflagration, February 2–3, 1928.*
99. *Boston Globe*, "Boston Firemen Return to Station."
100. Belknap, "Fall River Has Another Big Fire," 144.
101. *Fall River Conflagration, February 2–3, 1928.*
102. *Boston Globe*, "Woman's Term Cut by Fall River Fire."
103. *Fall River Conflagration, February 2–3, 1928.*
104. *Fall River Herald News*, "Witnessed Great Fire."
105. *Fall River Conflagration, February 2–3, 1928.*
106. Ibid.

Chapter 5

107. Silvia, "Flint Affair," 343. The section on "the Flint affair" and the early French Canadian influx is from Silvia's unpublished dissertation.
108. Silvia, "Spindle City," 384.
109. Harrington, *Call to Save*, 30. All of Harrington's observations come from this source.
110. McFarland, "Master's Work Destroyed."
111. Ibid.

112. *Notre Dame de Lourdes Memorial Book*, September 1983.
113. Lachance, "Church Bells Weighing Tons."
114. *Notre Dame de Lourdes Memorial Book*, September 1983.
115. Ibid.
116. Ibid.
117. Richard and Franklin, "I'm Heartsick."
118. See Appendix, Bishops of the Roman Catholic Diocese of Fall River.
119. *Notre Dame de Lourdes Memorial Book*, September 1983.
120. Richard and Franklin, "I'm Heartsick."
121. Devitt, "Notre Dame Still Alive." All of Masse's observations come from this source.
122. Lachance, "Probe Underway."
123. *Fall River Herald News*, "More Than 26 Departments Responded to Call for Aid."
124. Harrington, *Call to Save*, 31.
125. Negri and Fox, "Fall River Loses Piece of History."
126. Garland, "Damage in Fall River Fire."
127. Rhimes, "Flint Fire Recounted." All of Shea's observations come from this source.
128. Fox, "Fall River Rallies."
129. Ibid.
130. Harrington, *Call to Save*, 33.
131. Lachance, "Church Bells Weighing Tons."
132. Fox, "Fall River Rallies."
133. Lachance, "Church Bells Weighing Tons."

BIBLIOGRAPHY

General History

Belanger, Marc N. *A Guide to Fall River's Mills and Other Industrial Sites*. N.p.: self-published, 2013.

Dunwell, Steve. *The Run of the Mill: A Pictorial Narrative of the Expansion, Domination, Decline and Enduring Impact of the New England Textile Industry*. Boston: D.R. Godine, 1978.

Earl, Henry H. *A Centennial History of Fall River, Mass*. New York: Atlantic Publishing and Engraving Company, 1877.

Fall River and Its Manufactories, 1803–1878. 5th ed. Fall River, MA: Benjamin Earl & Son, 1878.

Fenner, Henry M. *History of Fall River*. New York: F.T. Smiley Publishing Company, 1906.

———. *History of Fall River, Massachusetts*. Fall River, MA: Munroe Press, 1911.

Haywood, Charles F. *General Alarm: A Dramatic Account of Fires and Fire-Fighting in America*. New York: Dodd, Mead, 1967.

Higgins, Alfred. *A Comprehensive Dictionary of Textile Terms*. Fall River, MA: Dover Press, 1948.

Hurd, Duane Hamilton. *History of Bristol County, Massachusetts: With Biographical Sketches of Many of its Pioneers and Prominent Men, Part 1*. N.p.: J.W. Lewis & Company, 1883.

Lima, Alfred J., Everett J. Castro and Kenneth M. Champlin. *A River and Its City: The Influence of the Quequechan River on the Development of Fall River, Massachusetts*. 2nd ed. Fall River, MA: PearTree Press, 2013.

Linton, George E. *Natural and Manmade Textile Fibers: Raw Material to Finished Product*. New York: Duell, Sloan and Pearce, 1966.

Lynch, Captain Thomas E. *History of the Fire Department of Fall River, Massachusetts*. Fall River, MA: Almy & Milne, Publishers, 1896.

Martins, Michael, and Dennis Binette. *Parallel Lives: A Social History of Lizzie A. Borden and Her Fall River*. Fall River, MA: Fall River Historical Society, 2011.

Peck, Frederick M., and Henry H. Earl. *Fall River and Its Industries*. Fall River, MA: Benjamin Earl & Son, 1877.

Phillips, Arthur Sherman. *The Phillips History of Fall River, Fascicle I*. Fall River, MA: Dover Press, 1944.

———. *The Phillips History of Fall River, Fascicle III*. Fall River, MA: Dover Press, 1946.

———. *The Phillips History of Fall River, Fascicle II*. Fall River, MA: Dover Press, 1945.

Silvia, Philip T., Jr. "The Spindle City: Labor, Politics, and Religion in Fall River, Massachusetts, 1870–1905." PhD diss., Fordham University, 1973.

Smith, Thomas Russell. *The Cotton Textile Industry of Fall River, Massachusetts: A Study of Industrial Localization*. New York: King's Crown Press, 1944.

Chapter 1: The Great Fire of 1843

Aurora Regional Fire Museum. "Handtub Fire Engine Historical and Technical Information." Handtub Expo. http://www.auroraregionalfiremuseum.org.

An Authentic Account of the Value of Property Destroyed by the Great Fire in Fall River, July 2, 1843. Boston: Dutton and Wentworth, Printers, 1844.

Brayton, Alice. *Life on the Stream*. Vol. 2. Newport, RI: Wilkinson Press, 1962.

Bronson, Reverend A. *An Address on the Anniversary of the Fire, Delivered to the Pearl St. Christian Chapel, July 2nd 1844*. Fall River, MA, 1844.

Dunbar, James N. "Disastrous City Mill Fire Bared Needs for Reforms." *Fall River Herald News*, September 23, 1977.

Fall River All Sorts. "Great Fire in Fall River." July 11, 1843.

Flanigan, Marion. "Fire of 1843 Was the First Major Tragedy Here." *Fall River Herald News*, July 2, 1983.

Fowler, Reverend A.M. *History of Fall River, With Notices of Freetown and Tiverton*. Fall River, MA: Almy & Milne, Publishers, 1862.

BIBLIOGRAPHY

Chapter 2: The Granite Mill Fire of 1874

Blewett, Mary H. *Constant Turmoil: The Politics of Industrial Life in Nineteenth-Century New England*. Amherst: UMass Press, 2000.

Fall River Daily Herald. "The Granite Mill Fire." September 19, 1911.

Fall River Herald News. "Granite Mill Fire Resulted in New Laws, Regulations." March 29, 1983, D6.

Greenlees, Janet. *Female Labour Power: Women Workers Influence on Business Practices in the British and American Cotton Industries, 1780–1860*. Burlington, VT: Ashgate Publishing Company, 1977.

History of the Granite Mill Fire. Fall River, MA: Slade & Franklin, 1874.

Inventory Form, Massachusetts Historical Commission. Located at Fall River Historical Society, n.d.

Lawlor, David S. *The Life and Struggles of an Irish Boy in America, an Autobiography*. Newton, MA: Carroll Publishing Company, 1936.

Chapter 3: The Steiger Store Fire of 1916

Boston Daily Globe. "Fall River Swept by $1,500,000 Fire." February 16, 1916.

———. "For Better Fall River." February 17, 1916.

———. "Search Ruins for Belongings." February 18, 1916.

Fall River Daily Globe. "Cathedral Luckily Uninjured." February 16, 1916.

———. "Costly Fire at American Linen Mills." February 14, 1916.

———. "Elks' Head Lost." February 17, 1916.

———. "Ellis Goldberg Sent in the First Alarm." February 16, 1916.

———. "Fine Work by the Firemen of Other Cities." February 16, 1916.

———. "Letters in Box Near Stieger's All Consumed." February 17, 1916.

———. "Loss of a Million and a Half by Midnight Conflagration." February 16, 1916.

———. "Mr. Steiger Will Rebuild Very Soon." February 16, 1916.

———. "Notes and Incidents of the Fire." February 16, 1916.

———. "Safes in the Burned District Show Contents in Good Shape." February 18, 1916.

———. "Scenes and Incidents." February 17, 1916.

Fall River Evening Herald. "All Available Police Respond to Alarm." February 16, 1916.

———. "Cause of the Big Fire Not Yet Determined." February 17, 1916.

———. "The Fire Inquest." February 23, 1916.

———. "Firemen Still Work on Ruins, All Thought on Rebuilding." February 17, 1916.
———. "Fireproof Roofing." February 17, 1916.
———. "Fully 20,000 People Attracted to Fire Zone." February 23, 1916.
———. "Lenox Hotel Guests Have Narrow Escape." February 16, 1916.
———. "Loss Exceeds Million Dollars." February 16, 1916.
———. "Origin of Steiger Fire Is a Mystery." February 23, 1916.
———. "Our Opportunity." February 16, 1916.
———. "Pumping Station Plant Renders Good Service." February 16, 1916.
———. "Reflection Cause of Alarm from Box Twelve." February 16, 1916.
———. "Snow Saved Many Houses." February 16, 1916.
———. "South Main Street Is Swept by Blaze; 23 Buildings Lost." February 16, 1916.
———. "Steiger Going to Have Better Store." February 16, 1916.
———. "Steiger Watchman Makes Statement." February 17, 1916.
———. "Telephone Company Did Excellent Work." February 16, 1916.
———. "Thoughtful Merchants Provided for Firemen." February 16, 1916.
Fall River Fire. Fall River, MA: The City News Company, 1916.
Municipal Journal (Municipal Journal and Engineer). "Fall River Business Section Swept by Fire." March 2, 1916.
Phillips, Craig. "How Arson Investigation Has Changed." Independent Lens. http://www.pbs.org/independentlens/blog/how-arson-investigation-has-changed.
The Standard (weekly insurance newspaper). "Lessons of the Fall River Fire." February 19, 1916, 190–93.
———. "Principal New England Losses of the Week." February 19, 1916. 190-193.

Chapter 4: The Great Fire of 1928

Belknap, Harry. "Fall River Has Another Big Fire." *Fire Engineering*, February 22, 1928.
Boston Globe. "Billboards to Proclaim Fall River's Optimism." February 4, 1928.
———. "Blast Levels Wall of Fire-Swept Mill." February 13, 1928.
———. "Boston Firemen Return to Station." February 4, 1928.
———. "Fall River Begins Rehabilitation Plan." February 4, 1928.
———. "Fall River Boiling All Drinking Water." February 4, 1928.
———. "Fall River Sets About Building." February 5, 1928.

———. "Guardsmen at Fall River Obey Rules Too Literally." February 6, 1928.
———. "Mill Cat Found Alive in Ruin of Plant." February 7, 1928.
———. "Phone Service Quits in City at 11 O'Clock." February 3, 1928.
———. "Suit for $500,000 Filed in Taunton." April 3, 1928.
———. "Tax Returns Lost in Fall River Fire." February 4, 1928.
———. "3,000,000 Moved by Fall River Bank." February 11, 1928.
———. "Woman's Term Cut by Fall River Fire." March 25, 1928.
The Fall River Conflagration, February 2–3, 1928. Boston: National Fire Protection Association, 1928.
Fall River Daily Globe. "Cause of Fire Still Mystery." February 3, 1928.
———. "Chief Thankful for Few Injuries Listed." February 4, 1928.
———. "Ghost-Like Tower of Pocasset Mill Must Be Destroyed." February 3, 1928.
———. "Madam's Dummies Go to Bank but Meet Harsh Cops." February 3, 1928.
———. "Plants Were Owned by Elston Company." February 6, 1928.
———. "Spectators Think Dummies Are Bodies of Fire Victims." February 3, 1928.
Fall River Herald News. "Center of City Swept by Devastating Fire." February 3, 1928.
———. "Finds Incendiarism Charges Groundless." February 8, 1928.
———. "Her Jewelry Lost in Fire." February 4, 1928.
———. "1928 Fire Destroyed Much of Downtown." October 17, 1978.
———. "Odd Occurrence at Conflagration." September 19, 1958.
———. "Wax Figures Are Removed." February 3, 1928
———. "Witnessed Great Fire." February 2, 1984.
Fall River, Massachusetts, City Directory. Boston: Sampson & Murdock, 1928.
Maiocco, Carmen J. *The Granite Block: Downtown in Fall River in the Mid-Twentieth Century*. N.p.: self-published, 1994.
Pictures of the Worst Fire in Fall River's History. New Bedford, MA: George H. Reynolds, 1928.

Chapter 5: The Notre Dame de Lourdes Fire of 1982

Boston Globe. "Complaints Sought Against Company Over Fall River Blaze." December 7, 1982.
———. "Fall River Rising From Fire's Ashes." May 13, 1982.

BIBLIOGRAPHY

Devitt, Phil. "Notre Dame Still Alive in Parishioners Hearts 30 Years After Fire." *SouthcoastToday*, May 10, 2012.

Fall River Herald News. "Businesses Hit by Fire." May 12, 1982.

———. "Fire Toll." May 12, 1982.

———. "Louis Destremps Designed and Directed Construction of Notre Dame de Lourdes." May 18, 1982.

———. "More Than 26 Departments Responded to Call for Aid." May 12, 1982.

———. "A Tragic Loss." May 13, 1982.

Flanigan, Marion. "Dr. Philip Silvia, Historian, Reflects on the Times When Notre Dame was Established." *Fall River Herald News*, May 18, 1982.

———. "It Was Like a Death in the Family." *Fall River Herald News*, May 12, 1982.

Fox, Wendy. "Fall River Rallies to Help Victims of Fire That Leveled Homes, Church." *Boston Globe*, May 13, 1982.

Garland, Russell. "Damage in Fall River Fire Is Put at $50 Million as the Cleanup Effort Begins." *Providence Journal*, May 13, 1982.

Harrington, Thomas J. *A Call to Save: The Memoir of a Fire Chaplain*. New Bedford, MA: Spinner Publications, 2006.

Higgins, Richard. "Fall River Lost More Than a Church." *Boston Globe*, May 16, 1982.

History of the Catholic Church in New England. Boston: Hurd & Everts Co., 1899.

Lachance, Aime. "Church Bells Weighing Tons Melted in Fire." *Fall River Herald News*, June 18, 1982.

———. "Probe Underway." *Fall River Herald News*, May 15, 1982.

Lewiston Daily Sun. "Contractor Defends Torch Atop Burned Church." May 13, 1982.

Maiocco, Carmen J. *Up the Flint*. N.p.: self-published, n.d.

McFarland, Bruce. "Master's Work Destroyed in Three-Hour Inferno." *Fall River Herald News*, May 18, 1982.

Negri, Gloria, and Wendy Fox. "Fall River Loses Piece of History." *Boston Globe*, May 12, 1982.

Notre Dame de Lourdes Memorial Book. September 1983.

Rhimes, Fred. "Flint Fire Recounted." *Fall River Herald News*, August 12, 1982.

Richard, Ray, and James Franklin. "I'm Heartsick. This Is My Parish." *Boston Globe*, May 12, 1982.

Silvia, Philip T., Jr. "The Flint Affair" from "The Spindle City: Labor, Politics, and Religion in Fall River, Massachusetts, 1870–1905." PhD diss., Fordham University, 1973.

Postscript: The Kerr Mill Fire of 1987

Boylan, Kate. "Are Mill Buildings Infernos Waiting to Happen? Maybe." *Standard Times*, January 18, 1987.

———. "Fire Left Livelihoods of Hundreds in Ashes." *Standard Times*, January 13, 1987.

———. "Time and Fuel Were Critical." *Standard Times*, January 13, 1987.

Boyle, Kate. "Another Fire Is Burned into Girl's Memory." *Standard Times*, January 13, 1987.

Boyle, Maureen, and Mark Vosburgh. "Damage Is $10 Million; 700 Return to Homes." *Standard Times*, January 13, 1987.

———. "Fall River Will Get Fire Aid." *Standard Times*, January 14, 1987.

Brown, Curt. "Residents Near Fire Describe Inferno." *Fall River Herald News*, January 12, 1987.

Dunbar, James N. "Kerr Mill Memories Survive Terrible Fire." *Fall River Herald News*, January 18, 1987.

Durand, Kathleen. "Fire Cause Still Unknown." *Fall River Herald News*, January 12, 1987.

———. "Many Face Loss of Jobs." *Fall River Herald News*, January 13, 1987.

———. "This Is the Real Thing." *Fall River Herald News*, January 18, 1987.

———. "Workers Witness Nightmare." *Fall River Herald News*, January 18, 1987.

Fall River Herald News. "Kerr Mill Conflagration." January 18, 1987.

Flynn, Sean. "City Council Asks State Aid to Meet Kerr Mills Fire Cost." *Fall River Herald News*, January 21, 1987.

———. "Remaining Mill at Kerr Mills Complex to be Razed." *Fall River Herald News*, January 27, 1987.

Jagolinzer, Bob. "City Seeks $1 Million in Federal Aid for Businesses Hurt by Kerr Mills Fire." *Providence Journal*, January 29, 1987.

———. "Fall River Asks State for Aid on Mill Fire." *Providence Journal*, January 21, 1987.

———. "Federal Agents Join State, City in Probe of Fire at Kerr Mills." *Providence Journal*, January 16, 1987.

Malinowski, W. Zachary. "Blaze Closes Route 195 for 8 Hours." *Providence Journal-Bulletin*, January 13, 1987.

———. "Origin of Kerr Mills Fire Traced to Boiler Room; Cause Unknown." *Providence Journal-Bulletin*, April 28, 1987.

McFarland, Bruce. "Loss from Fire May Reach $50 Million." *Fall River Herald News*, January 14, 1987.

———. "700 East End Residents Forced to Leave Homes." *Fall River Herald News*, January 13, 1987.

McFarland, Bruce, and Fred Rhines. "Hundreds Are Forced to Leave Their Homes." *Fall River Herald News*, January 12, 1987.

Pawlak-Seaman, Susan. "It Was 'A High Type of Operation.'" *Standard Times*, January 18, 1987.

Providence Journal. "Building Damaged in Mill Fire Must Fall After All, Official Says." January 27, 1987.

Rhines, Fred. "Overheated Pipe Started Kerr Fire." *Fall River Herald News*, April 28, 1987.

ABOUT THE AUTHOR

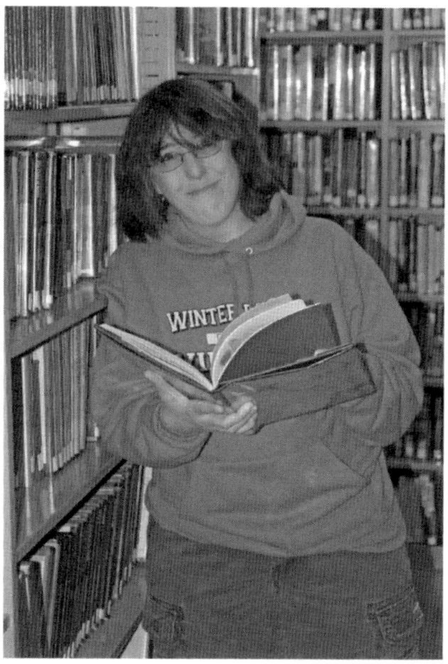

Stefani Koorey holds a doctorate in theater history and dramatic criticism and a master's of science in library science, digital libraries. She is the editor and publisher of *The Literary Hatchet* as well as the editor/owner of PearTree Press, specializing in Fall River history, the Borden murders of 1892, true crime, poetry and quirky fiction. With Al Lima, she is the co-founder of the Fall River History Club.

She has enjoyed a varied professional life as author, actress, director, stage manager, editor, professor, librarian, storyteller, box office manager, bibliographer, indexer, web designer and magazine publisher.

She is the author of *Fall River Revisited* (Arcadia Publishing, 2012); *Arthur Miller's Life and Literature: An Annotated and Comprehensive Guide* (Scarecrow Press, 2000); and *Story Programs: A Sourcebook of Materials*, 2nd edition (Scarecrow Press, 2000). She is also the coauthor of *The Preliminary Hearing in the Lizzie Borden Case, New Edition* (PearTree Press, 2009) and *The Trial of Lizzie Borden*, three volumes (PearTree Press, 2012).

ABOUT THE AUTHOR

In a strange turn of events, Stefani has become an expert on Lizzie Borden and the Borden murders of 1892 and actually lives in Fall River in a house that Lizzie owned before her death.